Exploration for Carbonate Petroleum Reservoirs

Exploration for Carbonate Petroleum Reservoirs

Elf-Aquitaine
Centres de Recherches de Boussens et de Pau

ANNE REECKMANN

GERALD M. FRIEDMAN

Rensselaer Center of Applied Geology
Affiliated with Brooklyn College,
City University of New York

A WILEY-INTERSCIENCE PUBLICATION

JOHN WILEY & SONS

New York • Chichester • Brisbane • Toronto • Singapore

Library of Congress Cataloging in Publication Data:

Main entry under title:

Exploration for carbonate petroleum reservoirs.
 Translation, expansion, and updating of: Essai
de caractérisation sédimentologique des dépôts
carbonatés. 2. Eléments d'interprétation.
 "A Wiley-Interscience publication."
 Includes bibliographies and index.
 1. Rocks, Carbonate. 2. Oil fields. I. Société
Nationale Elf-Aquitaine (Production). Centres de
Recherche de Boussens et de Pau. II. Reeckmann,
Anne. III. Friedman, Gerald M.

QE471.15.C3E78213   553.2′82   81-13144
ISBN 0-471-08603-7   AACR2

Printed in the United States of America

10 9 8 7

Preface

This book is an expanded version of a book on carbonate deposits originally published in French, Essai de Caractérisation Sédimentologique des Dépôts Carbonatés, Part 2 (1977). Unlike other books on carbonates, the original French book developed models based on the classic approach of Walther (1893–1894) and Lombard (1956). We liked this novel and creative approach and wanted to make it available to more readers. Hence we expanded and updated the French version and published it in English, adding some examples from North America.

This book is meant for professionals in the petroleum and mining industry, for students, and for academic researchers interested in carbonate rocks. We suggest that students or professional geologists who lack a background in carbonates read the relevant chapters of Friedman and Sanders (1978) to acquaint themselves with carbonate particles, processes of carbonate sedimentation, cementation, and the classification of carbonate rocks. This should provide sufficient knowledge for a beginner in carbonate studies to understand the level of treatment presented in this book. Those interested in additional and more specialized readings should follow up with Asquith (1979), Bathurst (1975), and Wilson (1975).

ANNE REECKMANN
GERALD M. FRIEDMAN

Troy, New York
December 1981

Acknowledgments

The original French version of this book was compiled by R. Cussey, J. Reulet, J. Bouroullec, R. Deloffre, R. Elloy, and M. Delmas with assistance from J. Aubert, H. Briotet, F. Calandra, O. Comby, G. Dailly, J. Etienne, M. Hamaoui, S. Jardine, D. Reyre, and L. Yapaudjian of Elf-Aquitaine, France. Helpful suggestions and constructive criticisms were made by R. G. C. Bathurst (University of Liverpool), A. V. Carozzi (University of Illinois, Urbana), and B. H. Purser (University of Paris). Photographic illustrations were by G. Poirel and figures by E. Labarthe.

For help with the English version of this book we thank Ron LeVan for preparing the figures, Mary Curl for compiling the index, Peter Kelliher for proofreading text and figures, and Stephen Scholle for proofreading galleys. Thanks are also extended to the Michigan Basin Geological Society for permission to reproduce figures. Sue Friedman helped in numerous ways behind the scenes and is gratefully acknowledged for her assistance. We especially thank Henri Oertli for his interest in seeing the book to completion.

S.A.R.
G.M.F.

Contents

1 DEPOSITIONAL ENVIRONMENTS AND GEOMETRY OF CARBONATE DEPOSITS   1

Depositional Environments, 1

Basis for Interpretation, 1
    Analytical Techniques, 1
    The Facies Model, 2

Major Kinds of Depositional Environments, 2
    The Nonmarine Environment, 3
    The Marine Environment, 3
    Shelf Environment, 3
    The Deep Ocean Basin, 6

Criteria for Characterizing Depositional Environments, 7
    Biological Criteria, 13
    Physical Criteria, 13
    Chemical and Climatic Criteria, 14

Vertical Relationships in Depositional Environments, 14

Transgressive-Regressive Sequences, 15

Upper and Lower Contacts, 16

Carbonate Sequences, 16

Geometry of Sedimentary Bodies, 19

Static Relationships, 19

Dynamic Relationships, 23

References, 37

Additional Reading, 37

2 CARBONATE RESERVOIR ROCKS   39

Methods of Study, 40

Petrophysical Analyses, 40
   Samples of Measurement, 40
   Units of Measurement, 40
   Methods of Measurement, 41

Petrographic Analysis, 41
   Macroscopic Analysis, 41
   Microscopic Analysis, 41
   Well Logging, 59
   Electric Logs, 59
   Radioactivity Logs, 59
   Productivity Measurements, 59

Kinds of Pores in Carbonate Reservoirs, 60

Classification of Porosity, 60

Primary Porosity, 60
   Framework Porosity, 60
   Intraparticle Porosity, 61
   Interparticle Porosity, 61
   Shelter Porosity, 61
   Breccia Porosity, 61
   Fenestral Porosity, 61

Secondary Porosity, 61
   Intercrystalline Porosity, 61
   Moldic Porosity, 62
   Vug and Channel Porosity, 62
   Fracture and Breccia Porosity, 62
   "Chalky" or Weathering Porosity, 63

Relationships Between Depositional Environments and Reservoirs, 63

Relationships Between Diagenesis and Reservoir Development, 65

Relationships Between Diagenetic Stages and Reservoir Development, 65
   Early Diagenetic Changes (Eogenesis), 69
   Stages of Burial Diagenesis (Mesogenesis), 74
   Late Subaerial Alteration (Telogenesis), 77

Pore Formation and Destruction as a Result of Diagenesis, 77
   Pore Formation, 77
   Inherited, Fossilized, and Stabilized Porosity, 77

Porosity Formed by Removal of Organic Matter, 77

Porosity Created by Organisms, 77

Porosity of Mechanical Origin, 78

Solution Porosity, 78

Pore Destruction, 78

Carbonate Cements, 78

Changes in Mineral Composition, 78

Cementation by Overgrowths, 78

Cementation by Noncarbonate Minerals, 79

Internal Sediment, 79

Pressure Solution, 79

Cementation by Recrystallization, 79

Conclusions, 79

References, 80

Additional Reading, 81

Appendix   Classification of Pores in Carbonate Rocks, 82

Classification Related to the Geometry of Pores, 83

The Classification of Waldschmidt et al. (1956), 83

The Classifications of Levorsen (1956), Harbaugh (1967), and
Klement (1971), 83

The Classification of Choquette and Pray (1970), 86

The Classification of Baillie and Vecsey (1972), 87

Classifications Based on Pore Interconnections, 87

Teodorovich (1943, 1958), 87

Classifications Based on the Relationships Between Rocks and Porosity, 94

Archie (1952), 94

Powers (1962), 94

Thomas (1962), 99

Sander (1967), 99

Ball (1968), 99

3   CASE HISTORIES OF CARBONATE RESERVOIRS   **100**

**Example 1   A Peritidal Dolomite Reservoir—The Ordovician Red River
Formation, Williston Basin, Montana, U.S.A., 101**

Geological Setting, 101

Facies and Environments, 103

Dolomitization, 103

Distribution of Porosity, 107
Conclusions, 107

**Example 2   A Dolomitic Peritidal Reservoir from the Jurassic
of Aquitaine, Southwest France, 107**

Data Provided by the Wells, 108

Conclusions, 111

**Example 3   High-Energy Shallow Marine Reservoirs in the Smackover
Formation (Jurassic), Gulf Coast, U.S.A., 113**

Geological Setting, 115

Vertical Sequence, 115

Horizontal Development of Lithologies, 117

Smackover Reservoirs, 117

Conclusions, 118

**Example 4   Carbonate Shelf Reservoirs: The Middle Jurassic of the
Paris Basin, France, 119**

Analysis and Interpretation of the Deposits, 120

Paleogeographic Synthesis, 127

Causes of Reservoir Deterioration, 127

Conclusions, 130

**Example 5   Diagenetic Reservoirs in Regressive-Transgressive Sequences,
Jurassic, Provence, Southeast France, 130**

Lithofacies, 130

Environmental Interpretations, 133

Vertical and Lateral Sequences, 134

Conclusions, 138

**Example 6   The Sligo Formation (Cretaceous) of the Gulf Coast, U.S.A:
Shelf-Edge Rudistid and Oöid Grainstone Reservoirs, 140**

Geological Setting, 141

Facies Development and Distribution, 142

The Black Lake Field, 143

Conclusions, 145

Example 7   Reef Reservoirs from the Middle Devonian of Northern Alberta, Canada, 147

Rainbow Reefs, Alberta, Canada, 147

Lithostratigraphy, 149

Diagenetic Alteration, 152

Conclusions, 155

Example 8   Pinnacle-Reef Reservoirs from the Middle Silurian of the Michigan Basin, U.S.A., 157

Geological Setting, 159

The Pinnacle Reefs, 161

Diagenesis and Reservoir Occurrence, 163

Conclusions, 165

Example 9   Carbonate Reservoirs in a Marine Shelf Sequence, Mishrif Formation, Cretaceous of the Middle East, 165

Analysis of the Mishrif, 165

Vertical- and Lateral-Sequence Development, 169

Mishrif Reservoirs, 170

Conclusions, 173

Example 10   Barrier-Reef and Carbonate-Shelf Sedimentation: Lennard Shelf, Devonian of Western Australia, 173

Geological Setting, 173

Sedimentary Bodies, 175

Facies and Depositional Environments, 176
   Shelf Environment, 179
   Shelf to Slope Transition, 180
   Slope and Deep-Basin Environment, 180
   Relationships Between Facies: The Sequences, 180

The Lennard Shelf Sequence, 183

Conclusions, 188

Example 11   The Permian of West Texas and New Mexico, U.S.A.; Multiple Reservoirs in Reef, Shelf, and Basin Settings, 190

Geological Setting and Carbonate Facies, 191

Reservoirs in the Permian Reef Complex, 194

Conclusions, 197

Example 12   A Carbonate Turbidite Reservoir—The "Scaglia Calcaire" (Cretaceous-Tertiary) of Central Italy, 198

Sedimentary Facies, 199

Lateral and Vertical Evolution, 199

Reservoirs, 204

Conclusions, 205

Example 13   Diagenetic Reservoirs Associated With Early Migration of Hydrocarbons, 206

References, 207

Index, 211

Exploration for Carbonate Petroleum Reservoirs

Depositional Environments and Geometry of Carbonate Deposits

DEPOSITIONAL ENVIRONMENTS

Depositional environments are natural geographic entities in which sediment accumulates. They are characterized by sets of biological, physical, and chemical parameters. The interaction of these parameters produces different sediment types or facies representative of different environmental conditions. A study of sedimentary facies in the rock record allows some interpretation of the conditions present in ancient depositional settings. In places, sets of environmental parameters are represented in the rock record only by a surface, for example, a bedding plane or disconformity surface. In most sequences these parameters correspond to a body or volume of sedimentary rock.

Basis for Interpretation

Many parameters characterize depositional environments, and these can be recognized through their effect on accumulating sediments. Environmental reconstruction is based on a knowledge of environmental processes and their products, which build up the sedimentary sequence. Facies models are used as a basis for understanding depositional environments and are constructed from real and theoretical studies, both of the rock record and of modern environments.

Analytical Techniques

The structural and textural details of each sedimentary layer, both on a large and a small scale, provide evidence for the reconstruction of the depositional environment. An ideal study would include the following:

1 Observation of the large-scale geometry of the sediment body in the field (sheet, lens, wedge, etc.) and establishment of its lateral and vertical relationships to adjacent bodies. Noted should be large- and small-scale internal structures.

2   Observation of the fabric, texture, and composition of the rock using polished slabs, thin sections, and acetate peels, both in the field and in the laboratory.

The scope of some studies is dictated by the kind of samples available (e.g., drill cuttings, cores, and sidewall cores), and there may be a lack of information about lateral extent, geometry, and the position of the sedimentary body in the sequence. Even so, when dealing with drill holes, a careful study of the logs can provide much information about the rocks in question, taking into account the effects of diagenesis, which are extremely important when dealing with carbonate rocks. Seismic studies are important for regional interpretation and provide valuable information about unit geometry, relationships, and extent.

In addition to the more common methods of study, detailed work on certain aspects of the rock can assist in interpretation: detailed micropaleontology, palynology, bulk-rock geochemistry, clay-mineral studies, organic geochemical studies, and scanning-electron microscopy.

The Facies Model

The basis for the study of sedimentary rocks and the best starting point for paleoenvironmental interpretation is a knowledge of modern environmental models (Heckel, 1972). Certain biological factors have changed with time, such as extinction and domination of certain genera and the environmental niche inhabited by certain genera; many factors have remained unchanged, however. The development of "ancient" models based on the rock record takes into account these factors and helps greatly in understanding the sedimentology of carbonates.

Major Kinds of Depositional Environments (Fig. 1-1)

Environments can be divided into two kinds—nonmarine and marine, the two being separated by the shoreline.

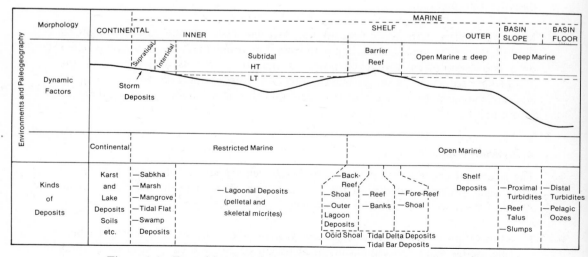

Figure 1-1 Depositional environments (HT = high tide, LT = low tide).

The Nonmarine Environment

Deposits of the nonmarine environment are generally very localized. These sediments are under the influence of meteoric water, and in carbonates this results in significant diagenetic changes. The principal kinds of nonmarine environments include deserts, glaciers, piedmonts, rivers, lakes, and caves. Carbonates are rare. Deposits of economic significance include alluvial placer deposits, peat, coal, marsh gas, phosphatic cave deposits, carbonate lake deposits, and mineralization associated with karsts.

The Marine Environment

The marine environment can be subdivided into the shelf, slope, and deep ocean basin. Depth of water on the shelf usually varies from 0 to 200 m. A break in slope separates the shelf edge from the deeper water of the ocean basin where depths commonly exceed 1000 m (Fig. 1-1). Profiles of the marine setting drawn normal to depositional strike may show considerable variation (Fig. 1-2).

Shelf Environment The environment of the shelf can be subdivided on the basis of a number of different characteristics: (1) morphology, (2) hydrodynamic regime, (3) salinity, and (4) light penetration (photic zone). The variations and interactions of these factors explain the diversity of carbonate deposits found in this environment, both in the geological past and at the present. Several models are necessary to cover the diversity of the shelf environment, models that allow, for example, for the presence or absence of a barrier or deep basin.

A barrier or shoal that is more or less continuous across the shelf is an important modifying feature (Fig. 1-3). The barrier itself can have complex and varied features, such as associated reef buildups, levees, deltas, dunes, barrier bars, tidal flats, channels, lagoons, and areas where active erosion occurs, shedding material into the surrounding environments. The relief of the barrier directly influences the energy, temperature, and chemistry (oxygenation and salinity) of the surrounding water, which in turn affects the biological activity on either side of the barrier (Fig. 1-3*a*). These variations give rise to the distinction between the inner- and outer-shelf environment (Fig. 1-3*b*).

Physical factors that influence the grain size, distribution, and grain shape of sedi-

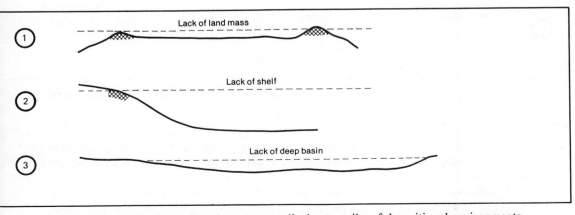

Figure 1-2 Schematic profiles drawn perpendicular to strike of depositional environments. Land, shelf, or basin may be absent.

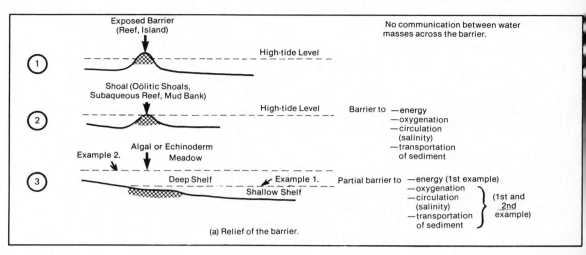

(a) Relief of the barrier.

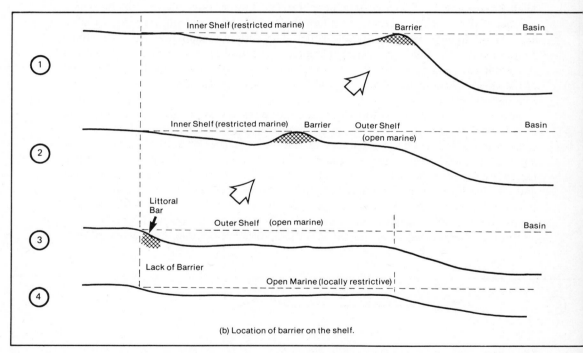

(b) Location of barrier on the shelf.

Figure 1-3 Schematic profiles drawn perpendicular to depositional strikes showing (*a*) variety in barrier morphology and (*b*) variation in barrier position.

mentary deposits in the shelf environment include waves, currents, swells, tidal currents, and wind. Variations in these factors determine the energy in the environment of deposition.

In epeiric (epicontinental) seas the tidal movement is dampened (Shaw, 1964; Friedman and Sanders, 1978, p. 360) (Fig. 1-4), but in shelf (pericontinental) seas, where tidal movement occurs, the inner shelf can be subdivided into a number of environments:

1 Supratidal environment, which is only rarely reached by high tides and storm surges.

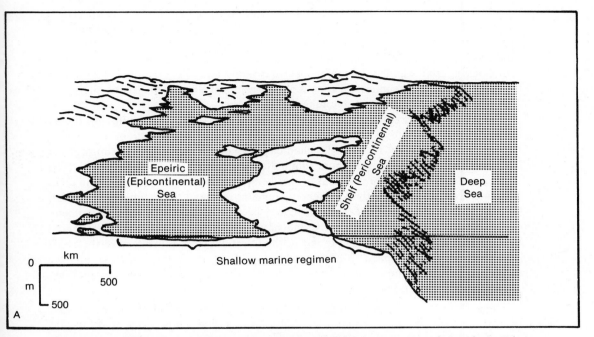

Figure 1-4 Epeiric and shelf seas showing their relationships to the continent (from Friedman and Sanders, 1978, p. 360, Figure 12-18A, modified after Heckel, 1972).

2   The intertidal environment, the zone between high and low water, where regular alternation between emergence and submergence occurs.

3   The subtidal environment, the zone that occurs below the lowest tides.

The term peritidal is used to encompass the sea-marginal environments, from subtidal to supratidal, which are subject to the effects of tidal fluctuation (Friedman and Sanders, 1978, p. 568).

In the rock record the tidal dynamics of an environment cannot be directly observed. Using facies association and the evolution of facies with time, it is often possible to deduce the approximate location of the shoreline at times in the geological past, and thus relate various parts of the sequence to environments established by using models of the present-day coast.

1   *Supratidal environment.*   Several subenvironments occur in this zone, including sabkhas, salt marshes, brine ponds, and coastal ponds. They develop above the normal influence of the sea and are only rarely flooded. The occurrence of well-developed deposits in this zone depends largely on the coastal profile. The nature of the deposits is greatly influenced by climate; for example, in an arid area sabkhas occur, while in moist climatic regions extensive salt marshes can develop.

The presence of both highly saline and fresh water in the supratidal environment makes it an important zone of early diagenetic alteration.

2   *Intertidal environment.*   Rhythmic sedimentation characterizes the intertidal environment, which is a zone of periodic emergence and submergence. It is a zone of abundant life, but living conditions are extreme and biota must be able to adapt to al-

ternating emergence and submergence, as well as to large variations in temperature, salinity, pH, and water chemistry. Climate has an important influence on intertidal deposits; for example, algal mats and their sedimentary products develop in the intertidal zone in arid climates. Subenvironments of the intertidal zone include the foreshore, beach, tidal channels, levees, mangrove swamps, and beach ridges. Together with the supratidal zone, the intertidal zone is an environment of early diagenetic alteration. This includes the formation of dolomite and evaporites.

The intertidal environment is generally a zone of high-energy, depending on the influence of tides, wind direction, currents, and the presence or absence of an offshore barrier or bar. In many respects the characteristics of the intertidal environment are indistinguishable from the near-shore subtidal environment.

3   *Subtidal environment.*   This is generally a low-energy environment, but in areas of high current and wave activity the energy remains high and the sediments are similar to those found in the intertidal zone. There may be growth of corals, oöid development, or the occurrence of channels, deltas, and bioclastic shoals. Deposits in the subtidal zone are not as variable as those found in the intertidal environment; nevertheless, it is an important environment of carbonate deposition. Microfauna is diverse, and the variations in water salinity, which depend mostly on water depth, produce a variety of sedimentary conditions.

High salinity, carbonate saturation, and poor oxygenation generally indicate a low-energy environment. These conditions result in a modified faunal association, usually one of low diversity. Recognition of such association of water conditions and fauna serves to distinguish a restricted marine environment (barred shelf or restricted basin) from an open marine environment (open shelf or basin). These distinctions are maintained for epeiric seas where no physical barrier to water circulation exists, but where the great width and shallow depth of the water body mean poor circulation, dampened waves, and low energy.

The depth of light penetration allows two zones to be distinguished in the subtidal environment: (1) the photic zone and (2) the aphotic zone. Biological conditions differ considerably between the two because photosynthesis cannot occur where no light penetrates.

It is clear that the many factors that affect marine shelf environments are interrelated. For example, the development of a topographic high across the shelf (e.g., an oöid shoal, barrier bar, or reef) affects the water chemistry of the inner shelf by restricting water movement (Fig. 1-3*b*). The same barrier acts as a damper on waves and currents, lowering the kinetic energy of the inner shelf environment. These effects are reflected in the kind of sediment that forms in the intertidal and supratidal environment. If the barrier is an algal buildup, a reef, or an active carbonate bank, its occurrence is restricted by the photic zone. Even in clear water this places it in depths that do not exceed a few tens of meters.

The Deep Ocean Basin   This is an environment where uniform and monotonous facies develop associated with long periods of nondeposition. The deep basin can be subdivided into a number of zones based on depth—bathyal (400–2000 m), abyssal (2000–6000 m), and hadal (deeper than 6000 m).

A slope environment can be recognized between the shelf and the deep basin. This is a zone of gravity-displacement products, which includes slumps, talus deposits, turbidites, and deep-sea fans.

Criteria for Characterizing Depositional Environments

Recognition of the various depositional environments is based on a number of observations.

1 Observation of macroscopic and microscopic features of the sediment. These include sedimentary structures, lithologies, mineral composition, and other aspects (Table 1-1).

2 Recognition of biological characteristics of the environment. These include the kind of fauna and flora present, the interaction between organisms and sediment, such as bioturbation, trapping, binding, and contribution of aragonite needles (Table 1-2).

3 Observation of physical and chemical conditions within the depositional environment, including salinity, dynamics, and climate (Tables 1-3 and 1-4).

These three features of the environment are interrelated, and all must be considered together. Fossil organisms and their traces should be among the factors related to biological activity, but they also have lithological implications. For example, lime mud can form by accumulation of skeletal debris (aragonite needles from algae), and laminated structures in lime mud can be the result of the actions of algal mats.

Based on the characteristics of the sediments alone, some physical and chemical conditions are difficult to infer; for example, temperature and depth. Some workers have tried to evaluate the absolute or relative depth of sediments at the time of deposition, but this inference is very difficult to make with any degree of accuracy and is often open to criticism. One of the most useful factors in the study of an environment is the energy of the environment at the time of deposition. In the rock record this can be determined from particle size, sorting, particle roundness, sedimentary structures, and fauna.

Climate often can be deduced from the occurrence and distribution of certain organisms and the presence or absence of evaporites and other deposits.

All the characteristics of sedimentary rocks must be carefully interpreted because many of the features seen in the rock record have been profoundly modified by diagenesis. The relative order of the diagenetic changes can sometimes be deduced, providing information about the rock from the time of its deposition to its occurrence at the present day. In many examples, distinguishing the origin of the initial sediment can be difficult. For example, micrite-like carbonate material in a limestone either has been originally precipitated directly from water, produced as lime mud by organisms and later lithified, or produced diagenetically as a cement (Friedman and Sanders, 1978, p. 175).

The occurrence of evaporite minerals depends on chemical and climatic factors. During early diagenesis, evaporites can easily be leached or replaced, thus removing valuable evidence from the deposit concerning the conditions of its origin. Likewise, specific diagenetic alterations shown by carbonate sediments and rocks can be related to a specific environment or environmental condition, such as dissolution, desiccation, brecciation, or precipitation of gravitational or meniscus cements (Dunham, 1971; Muller, 1971).

Table 1-1   Characteristics of Depositional Environments

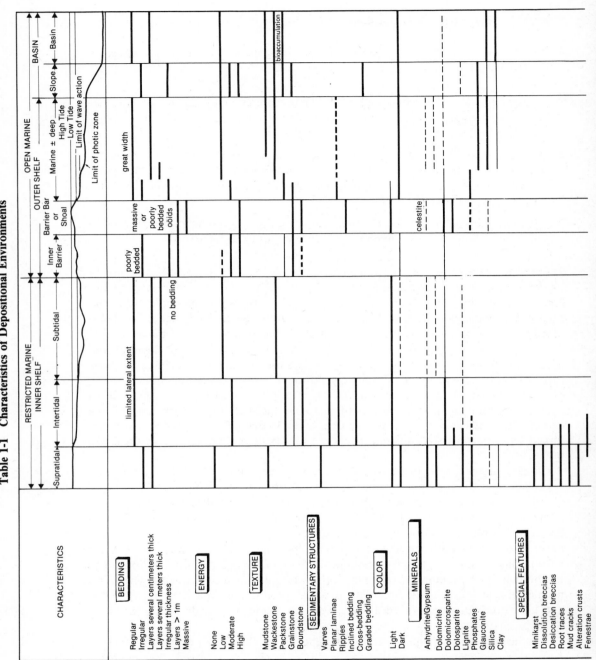

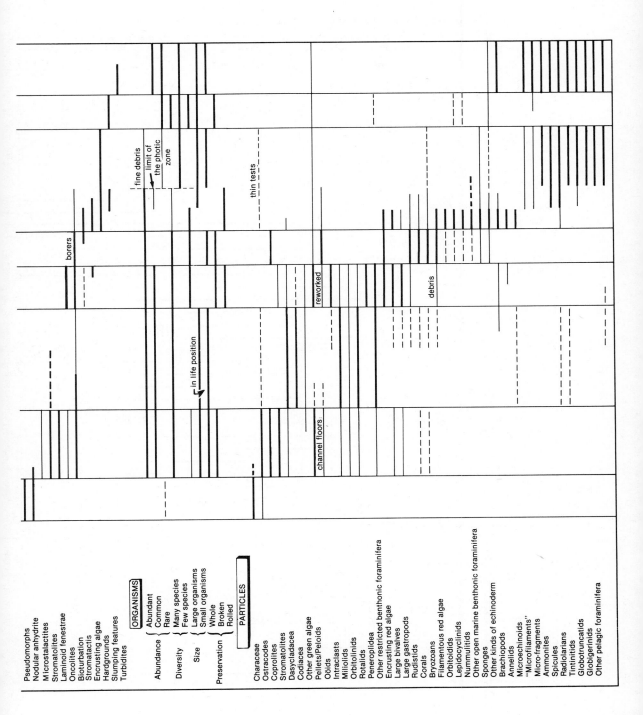

Table 1-2 Diagnostic Biological Characteristics of Depositional Environments

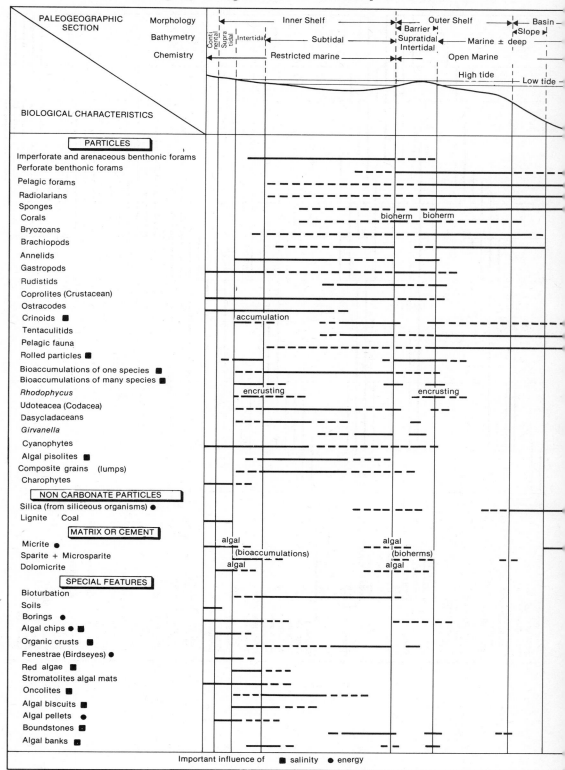

Table 1-3   Diagnostic Physical Characteristics of Depoistional Environments

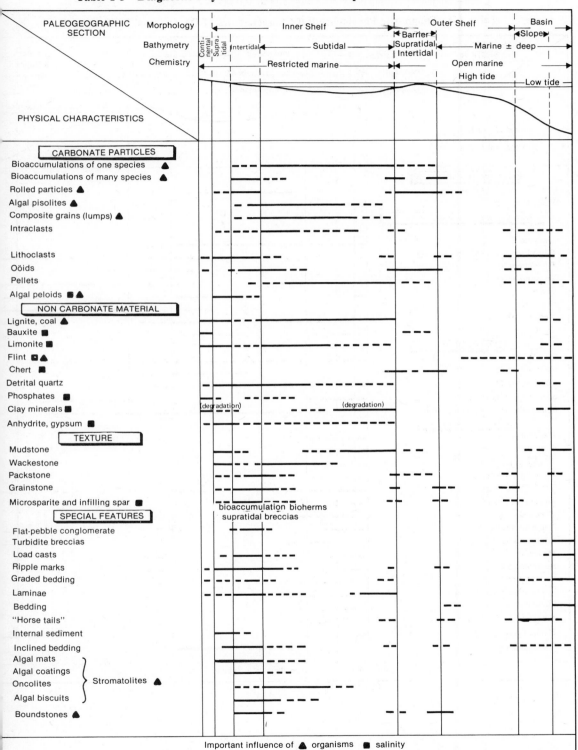

Important influence of ▲ organisms ■ salinity

Table 1-4 Diagnostic Physiochemical Characteristics of Depositional Environments

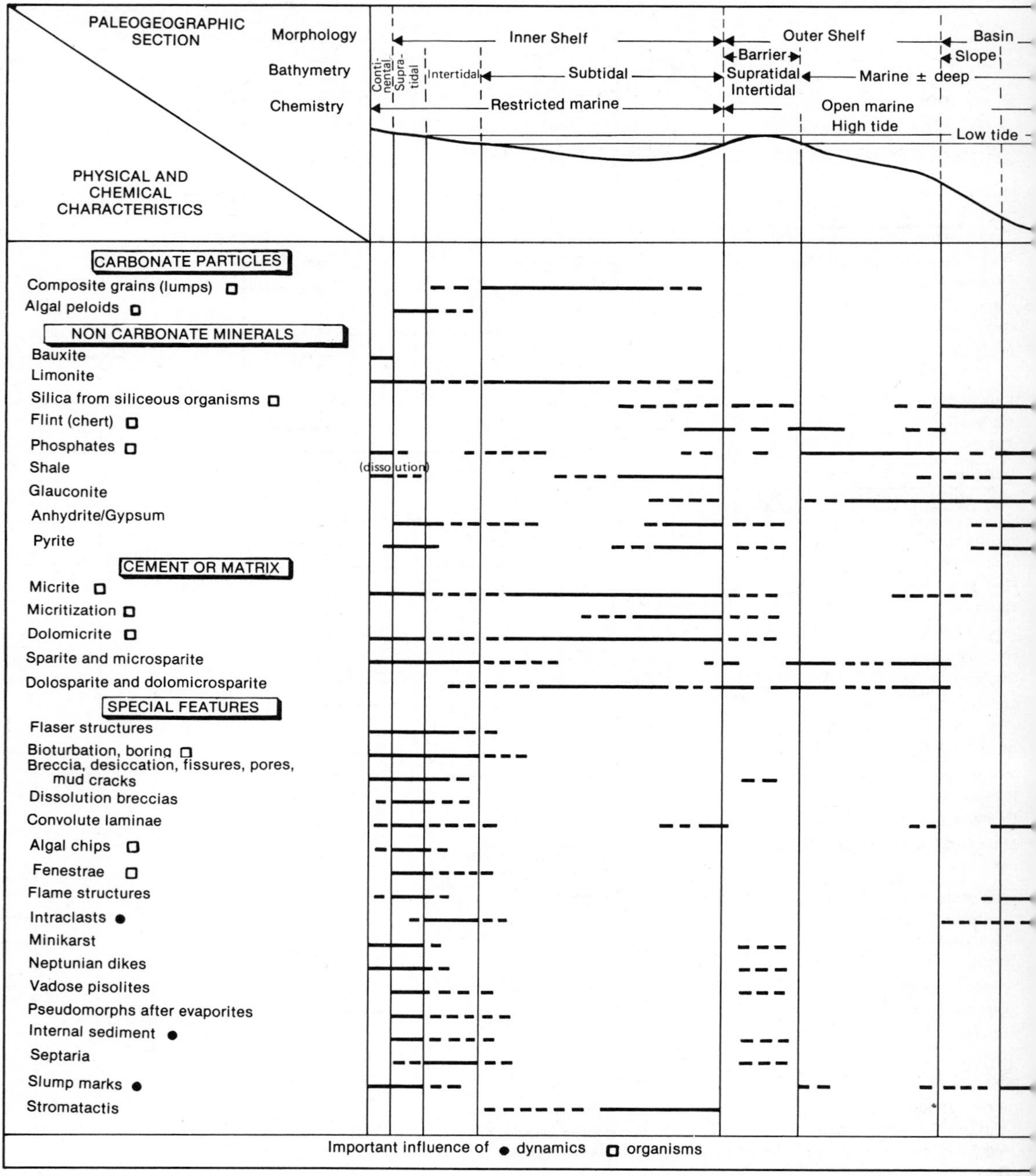

Column headers (PALEOGEOGRAPHIC SECTION):

- Morphology: Inner Shelf | Outer Shelf | Basin
- Bathymetry: Continental | Supratidal | Intertidal | Subtidal | Barrier | Supratidal/Intertidal | Marine ± deep | Slope
- Chemistry: Restricted marine | Open marine | High tide | Low tide

PHYSICAL AND CHEMICAL CHARACTERISTICS

CARBONATE PARTICLES
- Composite grains (lumps) □
- Algal peloids □

NON CARBONATE MINERALS
- Bauxite
- Limonite
- Silica from siliceous organisms □
- Flint (chert) □
- Phosphates □
- Shale   (dissolution)
- Glauconite
- Anhydrite/Gypsum
- Pyrite

CEMENT OR MATRIX
- Micrite □
- Micritization □
- Dolomicrite □
- Sparite and microsparite
- Dolosparite and dolomicrosparite

SPECIAL FEATURES
- Flaser structures
- Bioturbation, boring □
- Breccia, desiccation, fissures, pores, mud cracks
- Dissolution breccias
- Convolute laminae
- Algal chips □
- Fenestrae □
- Flame structures
- Intraclasts ●
- Minikarst
- Neptunian dikes
- Vadose pisolites
- Pseudomorphs after evaporites
- Internal sediment ●
- Septaria
- Slump marks ●
- Stromatactis

Important influence of ● dynamics □ organisms

Biological Criteria (*Table 1-2*)

Where the profile across the shelf includes a barrier, faunal characteristics are useful to distinguish the peritidal environment from the remainder of the profile. Because the barrier acts as a damper on water movement across the inner shelf, it is generally possible to distinguish the subtidal inner-shelf environment from the open marine environment of the outer shelf on the basis of fauna. The outer shelf is the environment of pelagic organisms such as planktonic foraminifera, and above-wave base colonial frame builders such as corals may be present.

If the profile is that of a simple sloping shelf without a barrier, the distinction between the inner and outer shelf may be difficult or impossible to make. The passage from one to the other is gradational, and the difference can most easily be related to the energy difference between shallow and deeper water. A second energy transition can be distinguished between the shelf and the deep-ocean basin.

In summary, the faunal criteria used to differentiate between the inner and outer shelf include:

1 The frequency and occurrence of low-diversity assemblages of benthonic organisms, which are more common in the inner-shelf environment.
2 The occurrence of algae, which are more abundant in the inner-shelf environment in shallow water where photosynthetic activity is at an optimum.
3 These observations must be used in conjunction with evidence from the sediments themselves.

Biological factors are themselves controlled by topography (depth), energy of the environment (which controls growth forms and species distribution), and climate (which controls faunal assemblage and nature of faunal hardparts).

Physical Criteria (*Table 1-3*)

Sediments from modern environments reflect the energy of deposition in their physical characteristics. These characteristics can be considered in two ways when dealing with the rock record.

1 The particles that constitute the deposit and have been carried into the depositional environment are considered in terms of their composition (quartz, mica, or skeletal fragments), size, degree of roundness, sorting, and the particle/lime mud ratio (i.e., whether grain supported or lime mud supported).
2 The deposit is considered in terms of its finest particles (the presence or absence of clay- or silt-size particles). This is a measure of the competency or of the smallest particle that has been able to resist reworking by currents and waves.

When dealing with carbonates rather than detrital sediments or rocks it is not always possible to relate directly the physical characteristics of the carbonate to the energy of the depositional environment. For example, in carbonates the size and shape of the grains are often a reflection of their origin alone and bear no relationship to the energy of the depositional environment; also, lime mud may be trapped in void spaces within a carbonate framework under high-energy conditions.

An estimate of the depositional energy of a carbonate sediment is best defined using the ratio between the matrix or micrite and the particles, in conjunction with the original morphological features of the particles. This allows a relative approximation of depositional energy, such as high, low, or moderate. It may be possible to infer that a deposit formed in a high-energy environment or a low-energy environment without having to relate it to depth. Dunham's classification (Dunham, 1962) is very useful for estimating depositional energy as long as the origin of the different kinds of particles in the deposit is considered; for example, particle shape and size may be related to biological origin, but micritization or later introduction of lime mud or cryptocrystalline cement into a boundstone or grainstone framework is likely to confuse interpretation of the depositional fabric.

When one is dealing with a complex shelf profile, the energy levels are complex and do not decrease uniformly with distance from the shore. The outer shelf seaward of a barrier is an area of high energy similar to the intertidal zone. Energy levels across the inner shelf may be low and comparable to those of the deep basin. A low-energy deposit can thus indicate deep water below wave base or shallow water in an area protected behind a barrier.

Chemical and Climatic Criteria (Table 1-4)

As part of a complex shelf profile the intertidal and supratidal zones have distinct chemical characteristics. Any distinction between chemical conditions in the subtidal inner shelf behind the barrier and the open marine outer shelf is difficult. Chemical factors are rarely independent of other criteria; for example, algal mats form in the supratidal zone in humid climates and in the intertidal zone in arid climates (Friedman et al., 1973). This development is controlled by the presence or absence of grazing and burrowing organisms (biological factor) and the distribution of these organisms depends on salinity (chemical factor), which in turn is controlled by the climate.

Climate also controls the kinds of chemical changes that occur during early diagenesis (Irwin, 1965), for example in a sabhka. Climate is very important, since it controls the growth of certain organisms, the chemistry of the water, and the kind of diagenetic overprint.

Evaluation of the various characteristics of carbonate rocks is important so that environments of deposition may be deduced and related to vertical sequence and to lateral and temporal relationships.

VERTICAL RELATIONSHIPS IN DEPOSITIONAL ENVIRONMENTS

In ancient sedimentary sequences the succession of lithologies indicates changes in the environment of deposition. These changes in lithology, which may be both lateral and vertical, define the boundaries of a particular sedimentary sequence in time and space. Each sequence is different and is defined by its lithology, faunal content, other geological characteristics, internal variations, and its limits. Its limits are sedimentological discontinuities of varying importance and duration. These discontinuities include simple bedding planes, scour surfaces, more prominent breaks in deposition as a result of erosion or sediment bypassing (diastems), hardgrounds (lithified surfaces that may show boring, encrusting, and erosion), and finally disconformities or angular unconformities between strata.

The scale at which the sequence is studied depends on the level at which it is sampled or observed. The degree of detail may vary from the level of individual beds to that of entire formations taken as a whole. This concept is important for the meaningful comparison of sequences, best done at the same level of detail for each sequence. However, the environmental changes that lead to a vertical succession of facies and the time intervals involved in these changes are quite variable. These differences result in several levels at which facies changes can be considered, and enforces the realization that within a sequence that shows change in facies with time there may be many smaller-scale variations.

Transgressive-Regressive Sequences

In detrital rocks a sequence can show a decrease in particle size toward the top (fining-upward or positive sequence) or a coarsening of particle size toward the top (coarsening-upward or negative sequence). These trends correspond to decreasing and increasing energy levels during deposition.

In carbonate rocks particle size variation with time depends as much on the kind of *in situ* organisms that are adding material to the sediment as on the particle-size distribution affected by the current. It is thus difficult to determine energy changes in carbonate environments using only the particle size of the deposit. However, carbonate sequences exist that, when related to the evolution of the depositional basin, do show fining-upward or coarsening-upward cycles similar to those shown by detrital deposits (Fig. 1-5). These cycles can be related to an overall increase or decrease in energy within the environment of deposition.

Sequences, which show vertical changes in particle size paralleling the general evolution of the sedimentary basin through regressive and transgressive cycles, may also show other features. In regressive sequences (offlap) the strata show a greater marine influence at the base while a transgressive sequence (onlap) shows the opposite trend.

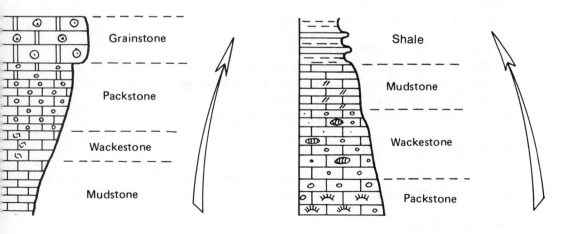

Coarsening-upward sequence
(energy increasing towards the top)

Fining-upward sequence
(energy decreasing towards the top)

Figure 1-5 Coarsening or negative and fining-upward or positive sequences.

To determine facies relationships and the correct order of vertical succession, a number of approaches may be taken:

1 Visual observation of the order in which strata succeed one another in the field or in the subsurface.
2 Statistical analysis of the strata in vertical sequence to test sequential or nonsequential facies relationships.
3 Arrangement of the strata in a logical order based on observations in modern environments and in models of ancient depositional environments.
4 Comparison of a completely developed sequence of sequential facies with partial or incomplete sequences.

Walther (1893–1894) and Lombard (1956) drew attention to the sequential analysis of deposits, no longer considering them as a vertical succession of lithologies, but rather as a continuous succession of units each to be studied in relation to those immediately above and below. Sequential analysis tries to group depositional facies or units into fundamental or ideal sequences that represent complete sedimentary cycles or conformable depositional sequences.

Upper and Lower Contacts

The base of a sedimentary sequence corresponds to the return of deposition as a result of:

1 A change in the depositional pattern, such as a decrease in kinetic energy.
2 Variations in eustatic or tectonic conditions that are accompanied by a sudden change in environment, hence a sedimentary unconformity.

The upper limit of a sequence generally corresponds to a break in sedimentation. In most depositional basins sedimentation occurs in small, localized areas and the time interval represented by deposition is much less than that corresponding to nondeposition. Breaks in deposition may result from:

1 Emergence because of basin infilling.
2 Change in the depositional environment, for example, introduction of a barrier to sedimentation, major changes in the current regime, reduction of sediment supply, or reduced biological or biochemical activity.
3 Emergence because of eustatic or tectonic changes in sea level.
4 Erosion.
5 Non-deposition.

The interplay between subsidence, sedimentation, and eustatic changes in sea level result in changing succession of vertical facies within any one environment (Fig. 1-6).

Carbonate Sequences

The development of sedimentary sequences is related to the environment of deposition and the kinetic energy of the environment at the time of deposition. Whether sequences

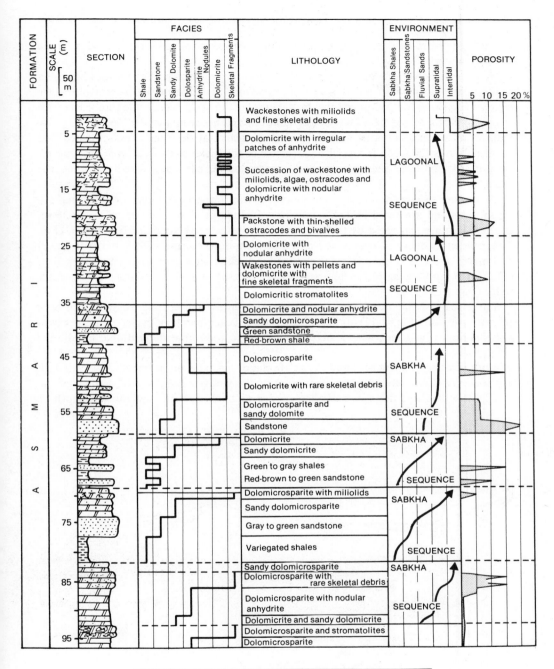

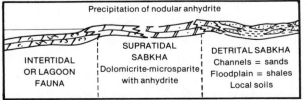

Figure 1-6 Example of a vertical succession of sequences in a shallow-water, arid depositional environment (Asmari Formation, Tertiary, Iran).

are fining-upward or coarsening-upward is primarily controlled by the evolution of the sedimentary basin.

Studies of accumulations of modern sediment have shown that the infilling of a sedimentary basin takes place as a build-out from the edge of the basin toward the center. The progradation of sedimentary deposits (Fig. 1-7) occurs in equilibrium with the marine profile (the base level of aggradation) and according to the biological, physical, chemical, and climatic conditions that are active at the time. The sequence that eventually fills the basin is a vertical sequence of the facies that occur in lateral succession at any one time and, as a result of progradation, become stacked one upon the other.

A number of common carbonate sequences are found in different depositional environments and are listed in Table 1-5. They include proximal and distal turbidite sequences that are generated by turbidity currents on the shelf slope and basin floor (Fig. 1-8), sequences of the littoral zone that grade from deep marine shelf deposits to beach or barrier deposits (Fig. 1-9), tidal channel and lagoonal sequences of the inner shelf (Fig. 1-10), and transgressive lagoonal sequences with abundant evidence of algal activity (Fig. 1-11).

Large-scale sedimentary sequences (megasequences) correspond to major periods of transgression (Fig. 1-12) or regression, whereas sedimentary cycles mark a transgressive-regressive event where there is a return to the initial depositional conditions (Fig. 1-13).

Both large-scale sequences and cycles are major episodes of sedimentation, although the sequences generated may be incomplete at the landward and seaward edges of the shelf. Close-to-shore deposits are often eroded or missing, while variation in depth or

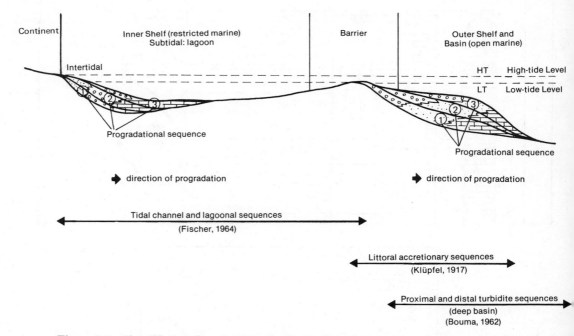

Figure 1-7   Simplified section showing the distribution of the main kinds of sequences in carbonate deposits.

Table 1-5 Common Carbonate Sequences and Their Depositional Environments

Kind of Sequence	Fig. No.	Environment	Vertical Sequence	Hydrodynamics
Distal turbidite	1-8a	Basin, slope of the outer shelf	Fining-upward (positive)	Turbidity current
Proximal turbidite	1-8b	Basin, slope of the outer shelf	Fining-upward (positive)	Mass-flow
Littoral sequence	1-9	Basin and/or outer shelf with a barrier (or beach)	Coarsening-upwards (negative)	Currents from waves, swells or tides
Tidal channel	1-10a	Inner shelf, supra-tidal barrier	Fining-upwards (positive)	Tidal currents
Lagoonal sequence	1-10b	Inner shelf, subtidal to supratidal	Fining or coarsening upwards	Intermittent currents
Transgressive lagoonal sequence with common algal laminites	1-11	Inner shelf and supratidal	(positive or negative)	Dominantly controlled by subsidence

change in depositional regime that has affected the shoreline sediments may not be recorded by the deeper-water facies. The development of these large-scale sedimentary sequences and cyclical deposits is controlled by the original morphology of the depositional basin, changes in plate tectonics or subsidence, supply of material, uplift, and eustacy. Factors involved in the generation of regressive or transgressive sequences and their effects are summarized in Table 1-6.

GEOMETRY OF SEDIMENTARY BODIES

Depositional environments have lateral and vertical extensions in time and space that are defined by the body of sediment they produce. The relationships and geometries of successive sedimentary strata mark the stages of sedimentological evolution of the basin. The geometry of carbonate sedimentary bodies thus enables reconstruction of basin evolution, both in *static* and *dynamic* terms.

Static Relationships

The geometry of sedimentary bodies is first considered in terms of simple relationships without taking into account any tectonic activity. Use is made of:

1 Field data from exposures and drill-hole sections which, when logically related, represent depositional environments that are predictable in horizontal and vertical directions (Walther's law).
2 Models of ancient and modern environments and their sedimentary sequences.

Simplified Section	Lithology	Special Features	Energy	Kinds of Particles	Environment
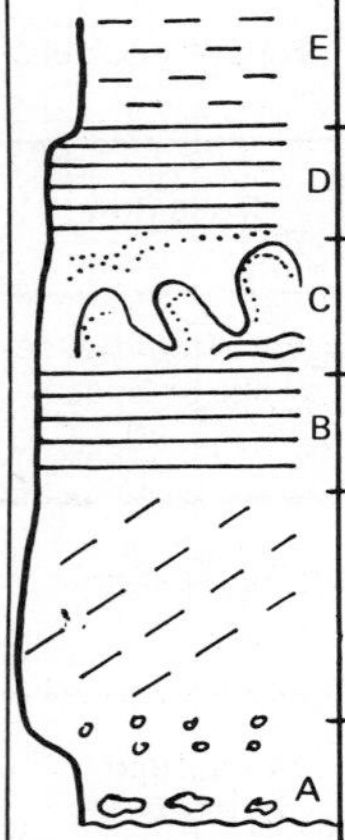 E	Argillaceous micrite	Mudstone	Low	Pelagic organisms	Basin
D	Micrite and micrite with gravel stringers	Horizontal laminae	Low	Gravel, peloids	or
C	Micrite ± argillaceous	Convolute laminae	Moderate	Gravel, benthonic and pelagic organisms	Outer
B	Micrite ± argillaceous	Horizontal laminae	Low		Shelf
	Cross-bedding	Submarine dunes (Hubert, 1964)	Moderate	Gravel intraclasts benthonic and	
A	Micrite and microsparite with shale pebbles	Graded bedding		pelagic organisms	

Thickness — 50 cm to 10's of meters

(a) Distal Turbidite Sequence (positive)

Simplified Section	Lithology	Special Features	Energy	Kinds of Particles	Environments
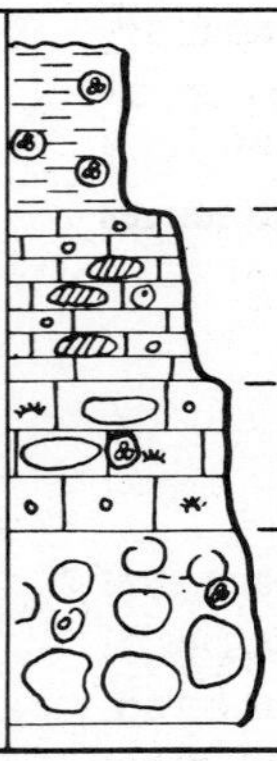	Shales with pelagic forams	Shale	Low	Pelagic organisms	Slope to Basin
	Wackestone with chert, micropellets and fine skeletal debris	Beds with chert	Low	Pelagic organisms, micropellets, fine echinoderm debris	
	Packstone with fine skeletal debris, peloids and pelagic forams	Microslumping	Moderate	Intraclasts, pelagic organisms, micropellets, microdebris	
	Packstone-grainstone with blocks and lithoclasts	Slumping and megaslumping	High	Pelagic and benthic organisms	

Thickness 1 to 5 m

(b) Proximal turbidite sequence (positive)

Figure 1-8 Modified Bouma sequences for carbonates. (*a*) Distal turbidite sequence. (*b*) Proximal turbidite sequence.

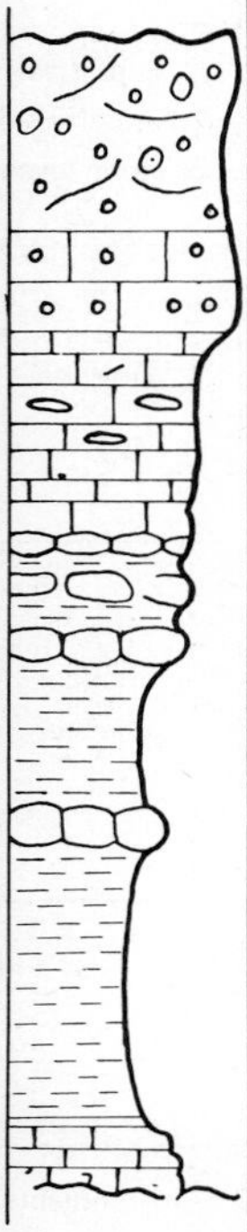

Simplified Section	Lithology	Special Features	Energy	Kinds of Particles	Environment	
	Light colored limestone (bedding cm. to dm. thick)	Boundstone and grainstone to packstone, local cross-bedding	High	Algae Corals Rolled skeletal debris Benthic fauna Oöids	Beach or Barrier Bar	
		Packstone-wackestone, graded bedding, slope breccias, chert	High to Moderate	Talus breccias Rolled skeletal debris Oöids Benthic fauna Peloids	Marine ± Deep to Deep Marine	
	Light to dark limestones, in places argillaceous (bedding dm. to m. thick), nodular bedding	Packstone-wackestone to mudstone, graded bedding, laminae	Moderate	Quartz Mica Lignite Glauconite Pyrite Rolled grains Bioturbation Benthic fauna Pelagic fauna Peloids Intraclasts	Marine ± Deep	Shelf
	Dark argillaceous limestones (bedding dm. to m. thick)	Mudstones	Low	Quartz Lignite Phosphates Pyrite Pelagic fauna Bioturbation Peloids	Deeper Marine	
	Glauconitic or ferruginous limestones	Wackestone to packstone with mineral crusts	Variable	Glauconite Lithoclasts Quartz Lithophages Phosphates	Hardground	

Figure 1-9 Accretionary sequence of the littoral zone.

By comparing the models and the field data the best reconstruction of sedimentary relationships is developed in stages as follows.

1 Sections are developed that still include some theoretical relationships and assumed distribution of deposits (Fig. 1-14).
2 Using knowledge of the depositional basin, a more realistic section is constructed that shows sediment distribution (Figs. 1-15 and 1-16).
3 A theoretical model of horizontal facies distribution is developed (Fig. 1-17).

The introduction of time enables reconstruction of paleogeographic maps for a particular time, T_1 (Fig. 1-18). A vertical succession of these maps at time T_1, T_2, and so on, allows reconstruction of a static block diagram that, depending on the availability of data, may be quite theoretical (Fig. 1-19) or close to reality (Figs. 1-20 and 1-21).

It is important to take into account models of modern depositional settings when developing environmental reconstructions from geological data (Fig. 1-22).

72100

Simplified Section	Lithology	Special Features	Energy	Kinds of Particles	Environment
Thickness <10 m	Dolomite and evaporites (early diagenetic)	Mudstone Boundstone Conglomerate Fenestrae Hardgrounds Mud cracks	Variable generally low	Quartz Mica Lignite Chert Pyrite Stromatolites Intraclasts Gravel (lumps) Lithoclasts	Supratidal (channel levees or storm levees)
	Dolomite and light colored limestones (beds dm. to m. thick)	Packstone Boundstone Cross-bedding Graded bedding Conglomerate Fenestrae Hardgrounds Mud cracks	Moderate	Quartz Mica Lignite Stromatolites Rolled skeletal debris Oncolites Intraclasts Peloids Oöids	Intertidal
	Light colored limestones (beds dm. to m. thick)	Boundstone Grainstone-packstone Cross-bedding	High	Algae Corals Rolled skeletal debris Benthic fauna Oöids Gravel (lumps)	Subtidal (channel

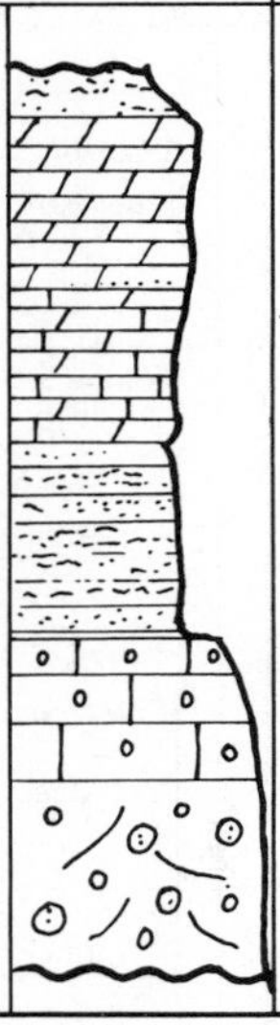

a) Tidal – channel sequence (positive)

Simplified Section	Lithology	Special Features	Energy	Kinds of Particles	Environments
Thickness < 10 m	Dolomicrite	Pseudomorphs after gypsum or anhydrite, laminae, leaching, vugs	None	Rare ostracodes Gastropods	Supratidal
	Algal Boundstone ± solution collapse and brecciation Algal boundstones	Solution breccias Boundstones	Moderate	Algal stromatolites Algal peloids Gastropods	High Intertidal to Intertidal
	Grainstone-packstone with benthonic foraminifera	Grainstone , Packstone	Moderate to High	Benthonic foraminifera Oncolites	Intertidal to Subtidal
	Wackestone	Wackestone	Low	Green algae	Subtidal

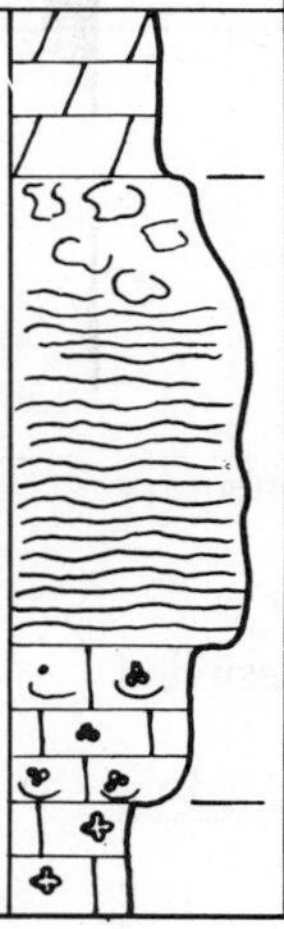

b) Lagoonal sequence (positive or negative)

Figure 1-10   Sequences for (*a*) tidal-channels and (*b*) lagoonal environments.

Simplified Section	Lithology	Special Features	Energy	Kinds of Particles	Environment
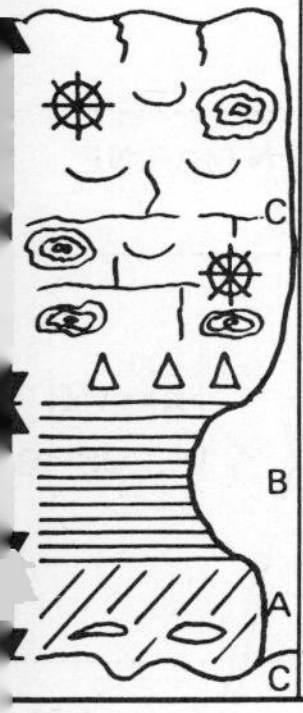	Thickly bedded microsparite and micrite	Cross-bedding (close to reef) Breccia of layer B	Moderate	Gravel (lumps) Oncolites Lithoclasts	Subtidal
B	Laminated dolomite	Stromatolites Mud cracks Fenestrae Desiccation	Low	Organic matter Lithoclasts	Intertidal
A	Green or red lag deposits	Breccia of layer C	Subaerial erosion	Iron Deposits	Subaerial
C					Subtidal

Figure 1-11 Transgressive lagoonal (peritidal) sequence with abundant algal laminae fenestrae.

Dynamic Relationships

While block diagrams showing the arrangements and evolution of depositional environments are useful interpretive tools, an understanding of basin dynamics is just as important. The infilling of sedimentary basins and sequential distribution of depositional sites depends on:

1 The initial morphology of the sedimentary basin.
2 The tectonic framework (deformation suffered by the basin, the plate tectonic setting that affects the lateral and vertical growth of the basin as a result of plate-margin movement and crustal subsidence).
3 Eustacy.
4 Supply of material to infill subsiding areas (rates of deposition relative to subsidence).

The relationship between the supply of material and the rate of subsidence is very important. In many basins a depocenter exists in which subsidence is at its maximum and the thickest deposits accumulate. However, the depocenter can shift with time (Fig. 1-23). A continuing balance between subsidence and deposition results in the buildup of a uniform deposit that has continued to form in an unchanging environment, there being no shoreward or seaward migration in environments as a result of changing sediment budget.

When the balance is upset, two kinds of variations in sequence can occur:

1 Where subsidence is slower than the rate of sedimentation, the basin gradually fills with sediment and becomes shallower (Fig. 1-24). When this happens, depositional facies shift toward the center of the basin.

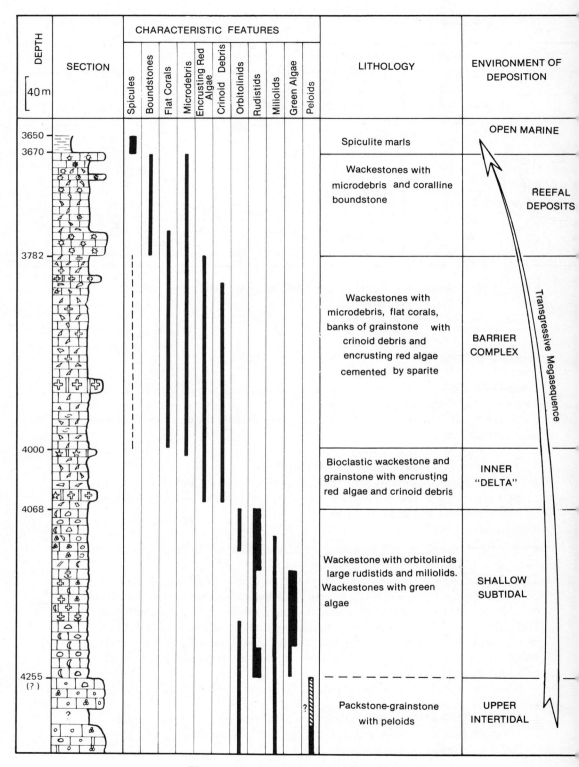

Figure 1-12 Example of a megasequence.

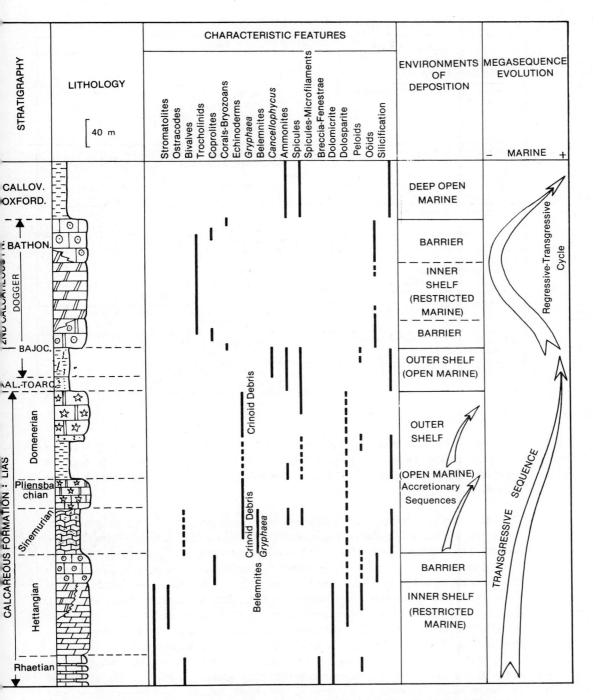

Figure 1-13 Example of a megasequence and sedimentary cycles (Jurassic of France).

Table 1-6   Factors Involved in the Generation of Regressive and Transgressive Sequences

REGRESSIVE SEQUENCES	
Controlling Factors (Acting alone or together)	Effects
1. Deposition more rapid than subsidence.	Regressive offlap with subaerial diagenesis of coastal deposits. Slumping occurs if the build-up of prograding sediment becomes too steep.
2. Uplift of basin margin.	Erosion and subaerial diagenesis of shoreline deposits.
3. Migration of depositional site into the basin.	Seaward displacement of deposition.
4. Eustatic fall in sealevel.	Erosion and subaerial diagenesis of coastal deposits.
TRANSGRESSIVE SEQUENCES	
Controlling Factors (Acting alone or together)	Effects
1. Rate of deposition slower than subsidence.	Marine transgression (onlap).
2. Subsidence at the basin margin.	Shoreward displacement of the site of deposition.
3. Migration of the depositional site towards the basin margin.	Shoreward shift in deposition.
4. Eustatic rise in sealevel.	Migration of depositional facies shoreward.

2   Where subsidence is more rapid than the rate of sedimentation, the basin becomes deeper closer to the shore (Fig. 1-24) and the facies shift shoreward. The deposits that form are different from those produced during the initial subsidence of the basin.

The change in basin shape with time is not only controlled by sediment progradation, but is also the result of erosion. The mobility of a basin, that is, the movement in the basement that controls the outline of the basin, is controlled by deep crustal tectonics or by isostacy. Tectonics are usually involved in the initiation of a basin (crustal subsidence creating a low in which sediment can accumulate, rifting and subduction of plates causing openings along continental margins). Tectonic movement may also mark the close of a basin's history when uplift carries it above base level.

The initial development of a sedimentary basin can be accentuated by down-to-basin step faulting. This provides additional volume for sediment entrapment.

Formation of a sedimentary basin—either clastic or carbonate—is very complex as a result of the many different mechanisms that may operate throughout the history of the basin. Such mechanisms as opening, subsidence, infilling, and uplift interact and change in importance with time. Some of the factors that operate are fundamental to basin formation, while others such as eustacy or climate simply modify depositional conditions.

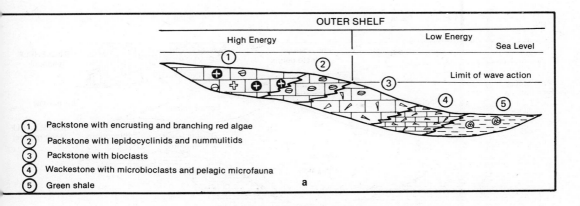

1. Packstone with encrusting and branching red algae
2. Packstone with lepidocyclinids and nummulitids
3. Packstone with bioclasts
4. Wackestone with microbioclasts and pelagic microfauna
5. Green shale

a

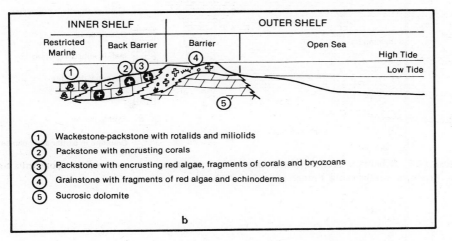

1. Wackestone-packstone with rotalids and miliolids
2. Packstone with encrusting corals
3. Packstone with encrusting red algae, fragments of corals and bryozoans
4. Grainstone with fragments of red algae and echinoderms
5. Sucrosic dolomite

b

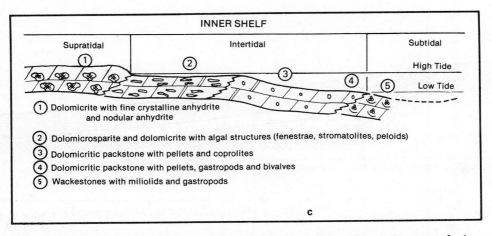

1. Dolomicrite with fine crystalline anhydrite and nodular anhydrite
2. Dolomicrosparite and dolomicrite with algal structures (fenestrae, stromatolites, peloids)
3. Dolomicritic packstone with pellets and coprolites
4. Dolomicritic packstone with pellets, gastropods and bivalves
5. Wackestones with miliolids and gastropods

c

Figure 1-14 Lateral sequences of deposits showing (*a*) accretion, (*b*) development of a barrier bar and (*c*) infilling.

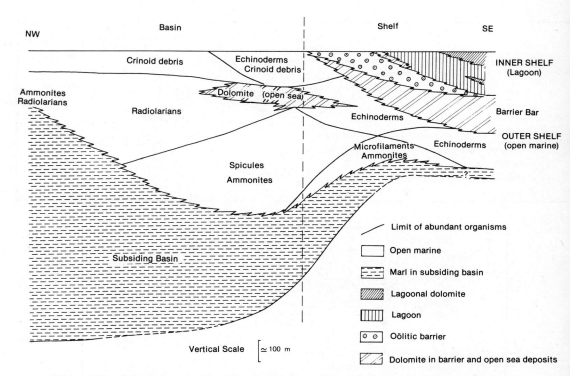

Figure 1-15 Schematic section of the distribution of Jurassic depositional environments and their products, southeastern France.

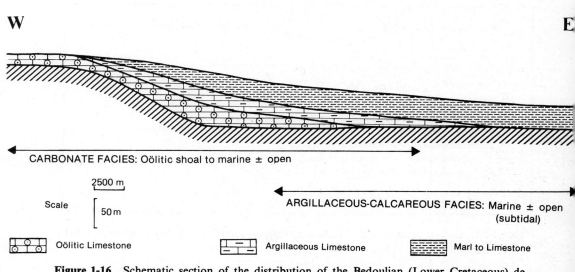

Figure 1-16 Schematic section of the distribution of the Bedoulian (Lower Cretaceous) deposits of Aquitaine (southwestern France).

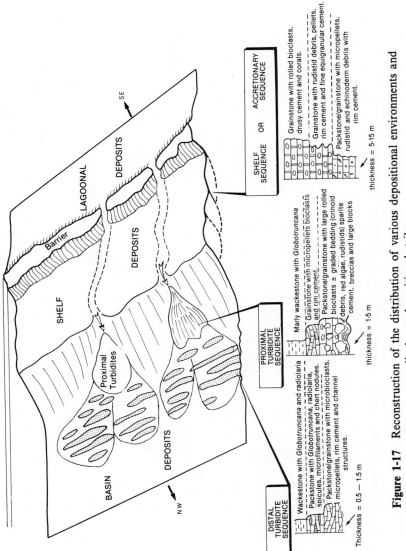
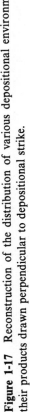

Figure 1-17 Reconstruction of the distribution of various depositional environments and their products drawn perpendicular to depositional strike.

29

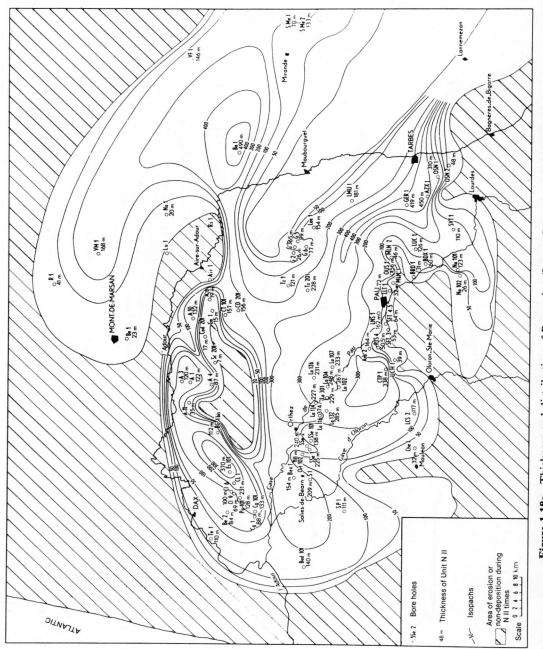

Figure 1-18a Thickness and distribution of Barremian (Lower Cretaceous) facies in south-western France.

30

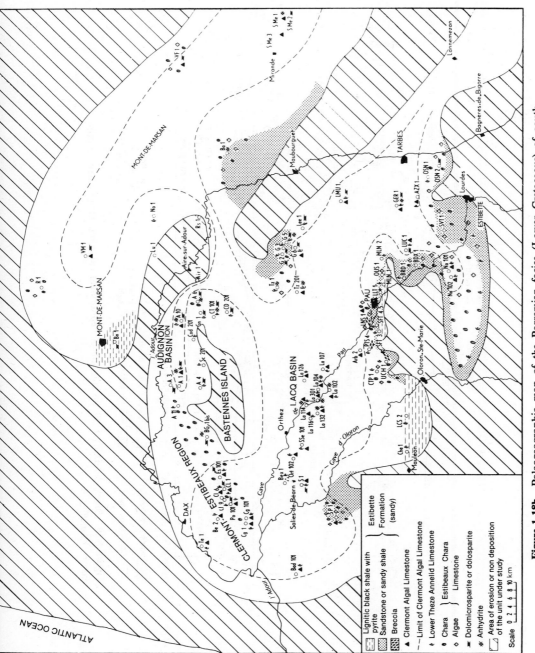

Figure 1-18b Paleogeographic map of the Barremian facies (Lower Cretaceous) of south-western France.

31

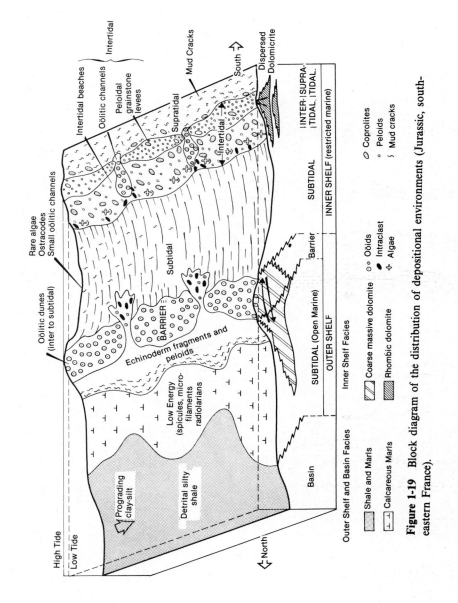

Figure 1-19 Block diagram of the distribution of depositional environments (Jurassic, southeastern France).

32

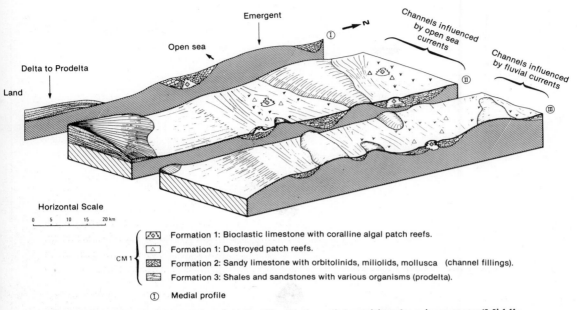

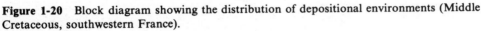

Figure 1-20 Block diagram showing the distribution of depositional environments (Middle Cretaceous, southwestern France).

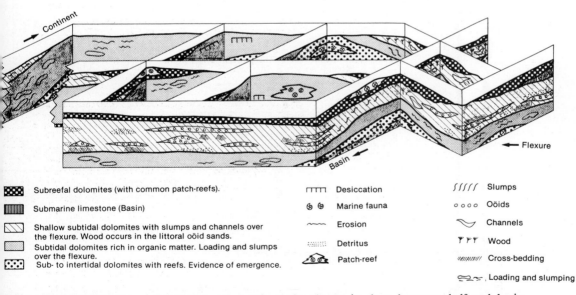

Subreefal dolomites (with common patch-reefs).

Submarine limestone (Basin)

Shallow subtidal dolomites with slumps and channels over the flexure. Wood occurs in the littoral oöid sands.

Subtidal dolomites rich in organic matter. Loading and slumps over the flexure.

Sub- to intertidal dolomites with reefs. Evidence of emergence.

⊓⊓⊓ Desiccation

☻ ☻ Marine fauna

∼∼ Erosion

⋯⋯ Detritus

🐚☻ Patch-reef

∫∫∫∫ Slumps

o o o o Oöids

∽ Channels

Ƴ Ƴ Ƴ Wood

〰〰 Cross-bedding

Loading and slumping

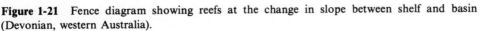

Figure 1-21 Fence diagram showing reefs at the change in slope between shelf and basin (Devonian, western Australia).

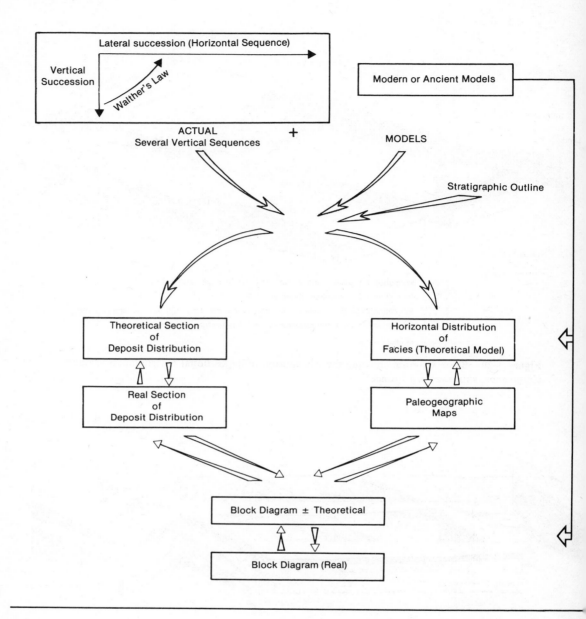

ACTUAL
Several Vertical Sequences

MODELS

Stratigraphic Outline

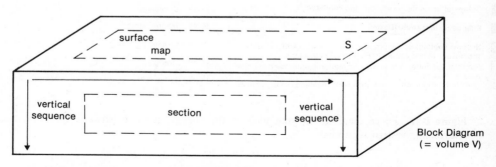

Figure 1-22   Stages in the three-dimensional reconstruction of basinal deposits.

34

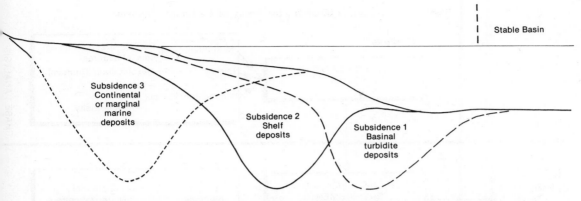

Figure 1-23 Relationships between the position of the axis of subsidence and the kinds of sedimentation.

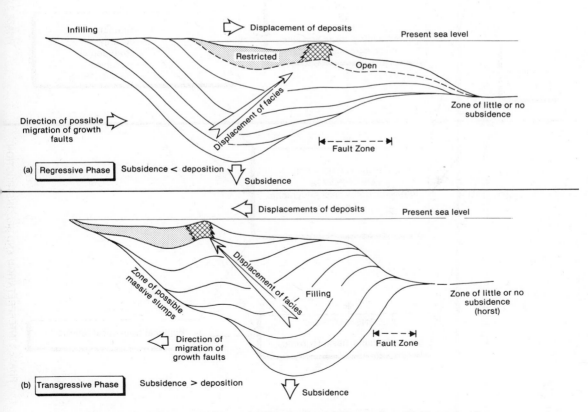

Figure 1-24 Sequences in a subsiding basin. (*a*) Regressive. (*b*) Transgressive.

Table 1-7 General Scheme for the Study of Carbonate Deposits

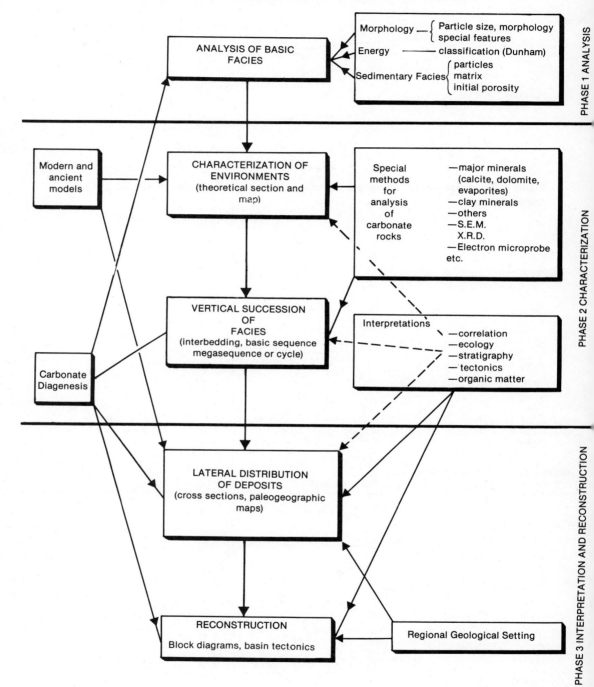

Carbonate deposits should be studied by comparing the characteristics of the sediments and the overall sedimentary sequence. The first approach identifies the environment in which the sediment was deposited; the second links together the sediments and the evolution of their geological setting with time. An approach similar to that shown in Table 1-7 makes interpretation of ancient sedimentary deposits much easier and more meaningful. It allows comparison of sedimentary rocks and their basinal setting with modern environments and the processes that are acting within these environments.

REFERENCES

Bouma, A. H., 1962, *Sedimentology of some flysh deposits. A graphic approach to facies interpretation:* Amsterdam, Elsevier Publishing Co., 198 p.

Dunham, R. J., 1962, Classification of carbonate rocks according to depositional texture, p. 108–121, in W. E. Ham, Ed., *Classification of carbonate rocks:* Tulsa, Okla., American Association of Petroleum Geologists, Mem. 1, 279 p.

Dunham, R. J., 1971, Meniscus cement, p. 297–300, in O. P. Bricker, Ed., *Carbonate cements:* Baltimore and London, The Johns Hopkins University Press, 376 p.

Fischer, A. G., 1964, The Lofer cyclothems of the Alpine Triassic, p. 107–149, in D. F. Merriam, Ed., *Symposium on cyclic sedimentation*, Kansas Geological Survey Bulletin 169, 2 vols., v. 1, 380 p.; v. 2, p. 381–636.

Friedman, G. M., and Sanders, J. E., 1978, *Principles of sedimentology:* New York, John Wiley & Sons, 792 p.

Friedman, G. M., Amiel, A. J., Braun, Moshe, and Miller, D. S., 1973, Generation of carbonate particles and laminites in algal mats—example from sea-marginal hypersaline pool, Gulf of Aqaba, Red Sea: *Bull. Am. Assoc. Pet. Geol.*, v. 57, p. 541–557.

Heckel, P. H., 1972, Recognition of ancient shallow marine environments, p. 226–286, in J. K. Rigby, and W. K. Hamblin, Eds., *Recognition of ancient sedimentary environments:* Tulsa, Okla., Society of Economic Paleontologists and Mineralogists, Spec. Pub. No. 16, 340 p.

Irwin, M. L., 1965, General theory of epeiric clear water sedimentation: *Bull. Am. Assoc. Pet. Geol.*, v. 49, p. 445–459.

Klüpfel, W., 1917, Über die Sedimente der Flachsee in Lothringer Jura: *Geol. Rundsch.*, v. 7, p. 97–107.

Lombard, A., 1956, *Geologie sedimentaire. Les series marines:* Paris, Masson et Cie., 722 p.

Müller, G., 1971, Gravitational cement, an indicator of the subaerial diagenetic environment, p. 32–35, in O. P. Bricker, Ed., *Carbonate cements:* Baltimore and London, The Johns Hopkins University Press, 376 p.

Shaw, A. B., 1964, *Time in stratigraphy:* New York, McGraw-Hill Book Co., 365 p.

Walther, J., 1893/1894, *Einleitung in die Geologie als historische Wissenschaft. Beobachtungen über die Bildung der Gesteine und ihrer organischen Einschlüsse:* Jena, Gustav Fischer, 1055 p. (3 vols.).

ADDITIONAL READING

Asquith, G. B., 1979, *Subsurface carbonate depositional models: a concise review:* Tulsa, Okla., The Petroleum Publishing Co., 121 p.

Burns, D. A., 1974, Changes in the carbonate component of recent sediments with depth: a guide to paleoenvironmental interpretation: *Mar. Geol.*, v. 16, p. M13–M19.

Chilingar, G. V., Bissell, H. J., and Fairbridge, R. W., Eds., 1967, *Carbonate rocks, origin, occurrence and classification:* Amsterdam, Elsevier Publishing Co., 471 p.

Coogan, A. H., 1969, Recent and ancient carbonate cyclic sequences, p. 5–16, in J. G. Elam and S. Chuber, Eds., *Symposium on cyclic sedimentation in the Permain Basin:* Midland, West Texas Geological Society, 203 p.

Friedman, G. M., Ed., 1969, *Depositional environments in carbonate rocks—a symposium:* Tulsa, Okla., Society of Economic Paleontologists and Mineralogists, Spec. Pub. No. 14, 209 p.

Friedman, G. M., 1972, Significance of Red Sea in problem of evaporites and basinal limestones: *Bull. Am. Assoc. Pet. Geol.*, v. 56, p. 1072–1086.

Horowitz, A. S., and Potter, P. E., 1971, *Introductory petrography of fossils:* New York, Springer-Verlag, 302 p.

Kinsman, D. J. J., 1969, Modes of formation, sedimentary associations, and diagnostic features of shallow-water and supratidal evaporites: *Bull. Am. Assoc. Pet. Geol.*, v. 53, p. 830–840.

Laporte, L. F., 1968, *Ancient environments:* Englewood Cliffs, N.J., Prentice-Hall, 116 p.

Reading, H. G., Ed., 1978, *Sedimentary environments and facies:* Oxford, Blackwell Scientific Publishers, 557 p.

Rigby, J. K., and Hamblin, W. M. K., Eds., 1972, *Recognition of ancient sedimentary environments:* Society of Economic Paleontologists and Mineralogists, Spec. Pub. No. 16, 304 p.

Selley, R. C., 1970, *Ancient sedimentary environments:* London, Chapman and Hall, 237 p.

Visher, G. S., 1965, Use of vertical profile in environmental reconstruction: *Bull. Am. Assoc. Pet. Geol.*, v. 49, p. 41–61.

Walker, R. G., Ed., 1979, *Facies models:* Geoscience Canada, Reprint Series 1, 211 p.

Wilson, J. L., 1975, *Carbonate facies in geologic history:* New York, Springer-Verlag, 471 p.

2

Carbonate Reservoir Rocks

Reservoir rocks are those in which oil, water, or gas fills pore spaces. The amount of porosity present in the rock is expressed as a percentage of the total volume of the rock. To make an economically viable producing unit, the pore spaces in a reservoir rock must be interconnected to allow movement and extraction of fluids. This connective network determines the permeability of the reservoir. Permeability is the capacity of porous material to transmit fluids and is related to the degree of interconnection of the pore network. The standard unit of measurement of permeability in rocks is the millidarcy.

Carbonate rocks and sandstones are the two common kinds of reservoir rock. Carbonate rocks may be characterized by high initial porosities linked to the high-energy levels in the environment of deposition. Initial carbonate porosity can greatly exceed that found in sandstones and may be as high as 60 to 80%, for example, in lime mud and reef frameworks. This original porosity is usually reduced by early diagenetic cementation, compaction, pressure solution, and late-stage cementation. In some reservoirs initial porosity is preserved or enhanced by dissolution involving both the original particles and any cement, resulting in so-called secondary porosity. As a function of diagenetic alteration, pore sizes in carbonate rocks may be independent of the size and packing of the original particles. Permeability does not depend on porosity or pore size, but is a function of the degree of pore interconnection. Important parameters that relate to permeability include pore-throat diameter, vertical and horizontal pore connection, and the presence of fracturing which can create an interconnected network independent of the original pore network.

Great heterogeneity is found in the "quality" of the pores and the petrophysical characteristics that are related to them. This heterogeneity occurs on all scales, from that of the large-scale fracture network throughout the rock strata to variation in micropores and their interconnections.

Research into the occurrence and formation of porosity is complex because of the many interrelated factors involved. Pore formation may be linked to deposition and to diagenesis in the original depositional environment as well as to later phases of diagenesis during burial and reexposure.

To find favorable zones of potential carbonate reservoir rocks in a sedimentary basin, the lateral and vertical evolution of the sequence involved must be known. The origin and types of porosity are important considerations because different basinal settings favor different kinds of pore formation. The mechanism of pore loss through compaction and diagenesis must also be considered, since such loss is more likely to occur in some settings than in others.

The description of kinds of pores, their sizes, and their geometries within rock strata should be combined with information about the environment, timing, and stage of pore formation and obliteration. Such information provides a basic framework within which the dynamic aspects of a reservoir can be studied.

Reservoir dynamics include the type and distribution of the network system that connects the pores. The relative ages of the pores and the interconnective network are also important when trying to date hydrocarbon emplacement and flushing. The study of modern carbonate deposition and diagenetic modification is the basis for understanding the formation of carbonate reservoirs. Diagenetic alteration is especially important because it can completely alter the initial character of a potential reservoir horizon.

METHODS OF STUDY

Reservoir evaluation includes both quantitative petrophysical measurements and qualitative or semiqualitative petrographic observations.

Petrophysical Analyses

Samples for Measurement

A plug, which is the most common kind of sample used for petrophysical measurements, is a piece of rock approximately 1 inch in length and diameter and usually cut from the center of a core. Plug samples are the most useful when dealing with homogeneous formations whose permeability, effective porosity, density, and so on, can be characterized by a small sample. Sidewall samples can also be used. These are small cores drilled or punched from the wall of the drill hole.

The "full diameter" is a sample taken from the whole core, usually 15 to 20 cm long. This type of sample is used for testing heterogeneous strata that cannot be properly represented by a small plug.

Units of Measurement

Porosity is expressed as a percentage of the total volume of the rock and varies in different directions through the rock (horizontal and vertical). The porosity of a rock indicates the amount of fluid or gas that a reservoir is able to contain. The effective porosity of a rock as opposed to total porosity represents only that pore space which is interconnected; however, the two values are usually very similar. Porosity can be estimated using a number of techniques, and porosity values vary according to the method of measurement. Most methods for measuring porosity are fluid injection techniques (LeRoy, LeRoy, and Raese, 1977).

Permeability of a rock is its capacity to transmit fluids. The basic unit of measurement is the millidarcy (one thousandth of a darcy). One darcy is the permeability that allows a fluid of 1 centipoise viscosity to flow at the rate of 1 cubic centimeter per second through a cross-sectional area of 1 square centimeter under a pressure gradient of 1 atmosphere per centimeter. The permeability of a rock also varies in the vertical and horizontal direction because of the relationship between permeability and rock fabric. The horizontal permeability is usually greater than the vertical permeability because the arrangement of particles and the partings in the rock tend to be parallel to bedding.

The wettability of a rock is a measurement that defines the affinity of a rock for oil and water. The degree of wettability is the angle that forms between the surface of the rock and the surface of contact with the water or oil. It is a measure of the ability of the liquid to form a coherent film on the surface of a rock as a result of the molecular attraction between the two. The wettability is an important factor in the removal of oil from a reservoir.

Methods of Measurement

In addition to the usual determination of porosity and permeability there are other tests that provide petrophysical information. Capillary pressure curves measured for a particular rock provide information about the efficiency with which oil can be removed from the pore system (Fig. 2-1). These curves may be generated using the mercury injection technique of Purcell (1949). The capillary pressure curves depend on the size of the throats connecting the pores and provide a measure of the pores in a reservoir rock connected by pore throats of sufficient size to enable oil to enter at a certain capillary pressure.

The porosity of a rock can also be determined using absorption of gamma rays. The absorption of the rays is a measure of the natural density of the rock, and from this value the porosity can be estimated, provided that the density of the rock is known.

Petrographic Analysis

Macroscopic Analysis

Large pores can be seen in hand samples with the unaided eye. This direct observation may provide some qualitative and quantitative data, but is limited by the nature of the sample and the inability of the observer to see very small pores. Direct observation is the best method for studying large pore and fissure systems, since an overall view often enables pore distribution patterns to be recognized.

More accurate information concerning pore geometry and arrangements is gained from the resin-injection methods (Wardlaw, 1976). These methods allow the intact pore network to be assessed as a whole. Lead injection of the pores enables X-ray examination of the sample to show the inner pore network. This technique reveals real porosity percentages and the exact geometry of the pores and their connective network.

Microscopic Analysis (Figs. 2-1 to 2-16)

Thin sections of rocks can be examined under the microscope, usually after the pore network has been impregnated with colored resin (Etienne and LeFournier, 1967). Quantitative estimations of pore percentages from thin section are not accurate because continuation of the pores in the third dimension is not evident. The thin-section sample is too small for large-scale features, such as fissures or solution cavities, to be readily recognized. Thin-section examination does allow the size of the

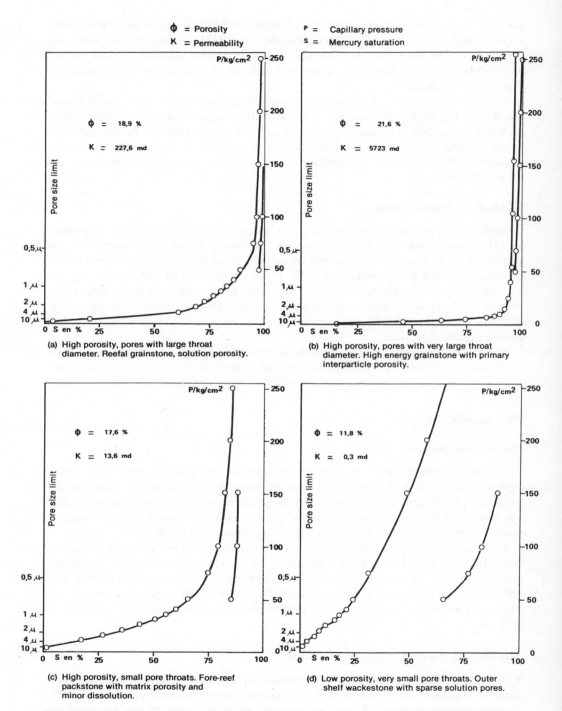

Figure 2-1 Capillary pressure or "Purcell" curves together with photomicrographs of the type of reservoir that they represent. (*a*) Reefal grainstone with solution porosity, (*b*) High-energy grainstone with interparticle porosity, (*c*) Fore-reef packstone with "matrix" or intra-micrite porosity, (*d*) Outer-shelf wackestone with solution pores.

42

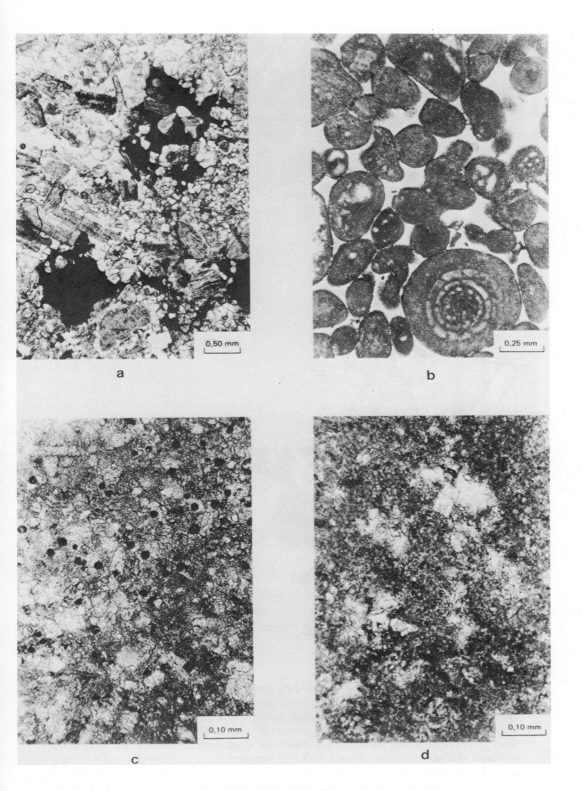

a

0,50 mm

b

0,25 mm

c

0,10 mm

d

0,10 mm

Figure 2-1 (*Continued*)

43

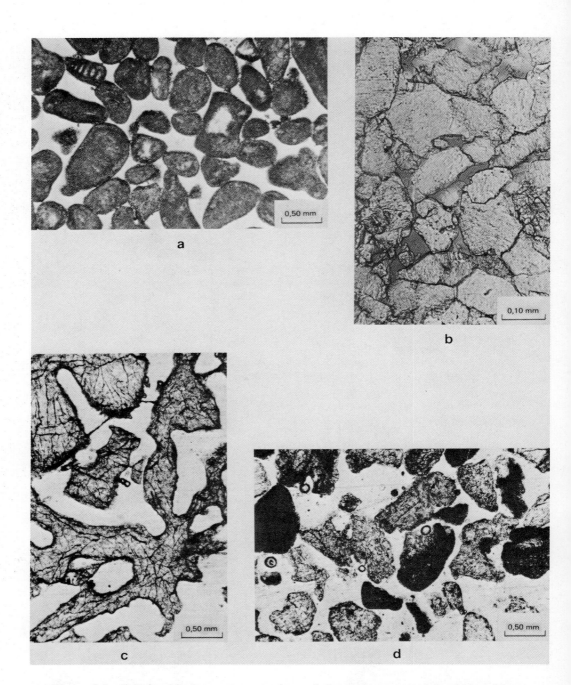

Figure 2-2 Relationship between pororosity and the environment of deposition (photomicrographs in plane-polarized light). (*a*) Interparticle porosity in a grainstone from a high-energy shoal environment. Particles include micritized bioclasts and sporadic oöids (pores are the white spaces). Dogger, Middle Jurassic of the Paris Basin. (*b*) Interparticle porosity (gray, resin-impregnated patches) in a grainstone slope deposit. Particles are recrystallized bioclasts. Turbidite breccia, Guinea Gulf. (*c*) Intraparticle or framework porosity within a coral builup. Quaternary reef, Pacific. (*d*) Interparticle porosity (light-colored irregular patches) preserved between bioclasts including echinoderm plates and algal fragments. Modern back-reef deposit, Pacific.

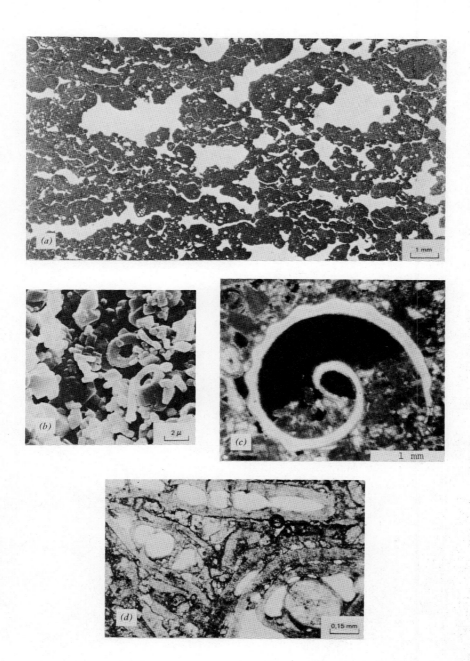

Figure 2-3 Relationship between porosity and environment of deposition (thin-section photomicrographs in plane-polarized light *a*, *c*, and *d*, scanning electron micrograph *b*). (*a*) Fenestral porosity (white patches) in an algal deposit. Particles include peloids (micritized bioclasts) and foraminifera. Moderate to high-energy shoal, Tunisia (*b*) Inter- and intra-coccolith porosity. Low-energy, open marine coccolith mud from the Cretaceous of the North Sea. (*c*) Shelter porosity within a gastropod, reef core composed of skeletal fragments, including coralline algae, in a cryptocrystalline cement. Modern reef Red Sea, Gulf of Elat (photograph G.M.F.). (*d*) Intraparticle porosity (light patches within nummulitids) in a nummulitid shoal deposit. Moderate-energy, Eocene of Gabes Gulf, Tunisia.

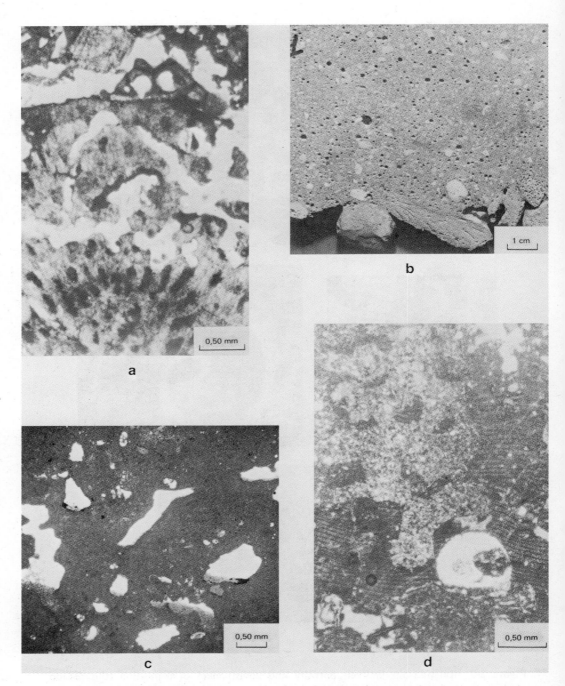

Figure 2-4 Relationships between porosity and early diagenesis—effect of organisms (thin sections in plane-polarized light *a*, *c*, and *d*; polished slab *b*. (*a*) Porosity formed by borings in a bivalve shell. Plio-Pleistocene of the Pacific. (*b*) Porosity formed by borings in a bioclastic limestone (dark circular patches). Quaternary from the Pacific. (*c*) Porosity formed by burrowers in a lime mud (light patches). Note the presence of internal sediment at the base of some of the pores. Low-energy deposit from the Dogger (Middle Jurassic) of the Paris Basin. (*d*) Filled burrows (light mottled patches) in red algae *Lithothamnium*. Plio-Pleistocene of the Pacific.

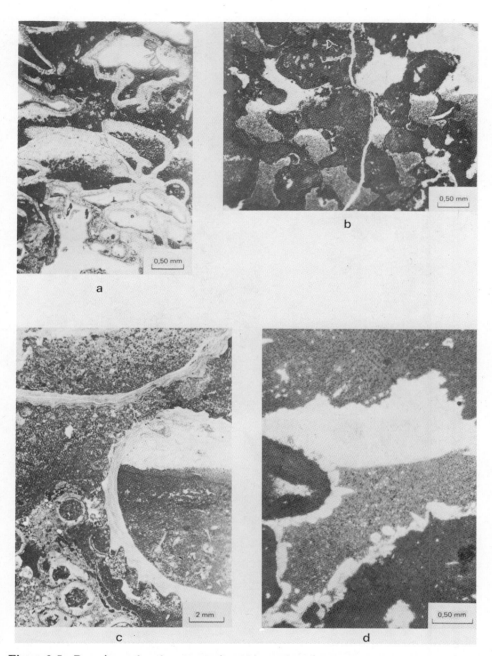

Figure 2-5 Porosity and early cementation (thin sections in plane-polarized light). (*a*) Intraparticle porosity within the chambers of foraminifera showing partial occlusion by peloidal micrite (an internal sediment cement) and a later sparry calcite mosaic (light area above the micrite). Packstone containing encrusting foraminifera, Mio-Pliocene of the Pacific. (*b*) Burrow porosity in an algal mat showing infilling by micritic internal sediment (base of the burrows) and calcite spar (light-colored patches). High-energy algal boundstone from the Dogger (Middle Jurassic) of the Paris Basin. (*c*) Intraparticle porosity within bivalve shells occluded by micritic internal sediment and coarse sparry calcite. Bivalve wackestone from the inner shelf, Lias of Morocco. (*d*) Vuggy porosity filled by dogtooth calcite, micritic internal sediment, and finally coarse sparry calcite. Tertiary mudstone, Italy.

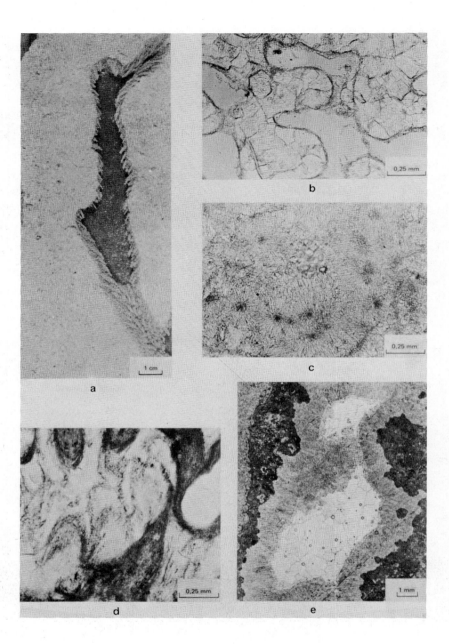

Figure 2-6 Porosity and early diagenetic behavior of aragonite (polished section *a*, photomicrographs in plane-polarized light *b*, *c*, *d*, and *e*). (*a*) Polished section of a coral showing almost complete dissolution of aragonite in the central zone forming a large void. Mio-Pliocene reef sediment from the Pacific. (*b*) Secondary porosity formed by dissolution of the aragonitic wall of a coral; the coarse spar fills the voids originally occupied by the soft parts of the coral. Mio-Pliocene reef sediment from the Pacific. () Intraparticle porosity filled by secondary aragonite needles. Coral from the Mio-Pliocene of the Pacific. (*d*) Alternation of aragonite (dark) into calcite (light) in a coral wall, reef sediment. Mio-Pliocene of the Pacfic. (*e*) Vuggy porosity occluded by fibrous aragonite (dark) and sparry calcite (light). Low-energy mudstone in a reworked Cretaceous deposit from Italy.

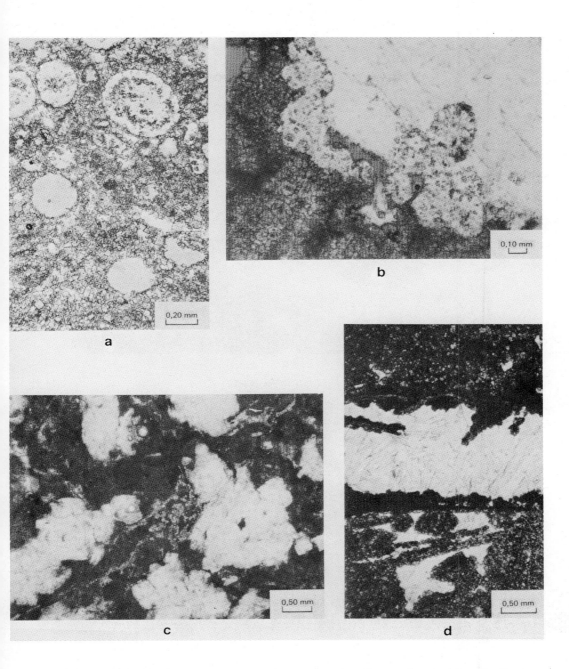

Figure 2-7 Relationship between porosity and the behavior of evaporites during early diagenesis (thin-section photomicrographs in plane-polarized light). (*a*) Secondary porosity formed by the dissolution of oöids and filled by anhydrite. Grainstone from an oölitic shoal, Permian of Iran. (*b*) Former vuggy pores now filled by anhydrite (light) in a dolomicrosparite. Dolomite crystals can be seen in anhydrite at the boundary between the two minerals. Inner-shelf deposit from the Tertiary of Iraq. (*c*) Replacement and pore-filling anhydrite developed within a dolomicrite. Low-energy, restricted marine sediment from the Tertiary of Iraq.(*d*) Radiating needles of anhydrite filling void space within and between particles and replacing matrix of dolomicrosparite. Low-energy deposit from a restricted marine environment, Tertiary of Iraq.

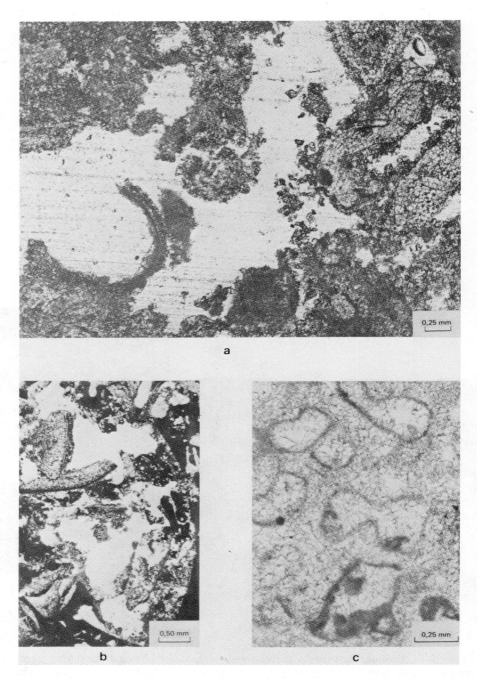

Figure 2-8 Relationship between porosity and early diagenetic dissolution/precipitation of calcite (thin-section photomicrographs in plane-polarized light). (*a*) Vuggy porosity formed by the dissolution of calcite in a slightly dolomitized packstone. Moderate-energy back-reef sediment from the Mio-Pliocene of the Pacific. (*b*) Vuggy porosity formed by the solution of calcite in a bioclastic packstone. Moderate-energy, open marine deposit from the Jurassic of southwestern France. (*c*) Porosity in a *Hexicoral* partially occluded by precipitation of calcite in the chambers. Alteration of the wall to calcite is also almost complete. Reefal boundstone from the Mio-Pliocene of the Pacific.

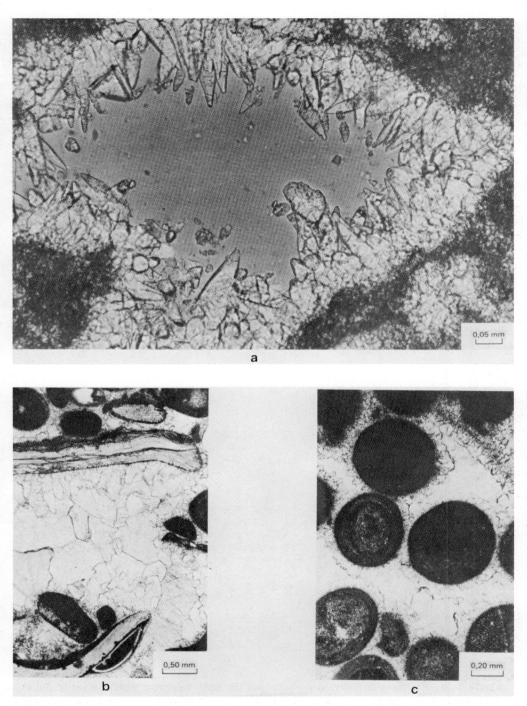

Figure 2-9 Relationship between porosity and early diagenetic cementation by calcite (thin-section photomicrographs in plane-polarized light). (*a*) Vuggy porosity partially occluded by void-rimming dogtooth calcite. Mio-Pliocene of the Pacific. (*b*) Interparticle porosity filled by a drusy calcite mosaic. Bioclastic grainstone from a shoal deposit. Jurassic of the Paris Basin. (*c*) Interparticle porosity partially filled by void-rimming calcite spar. Beachrock? from the Jurassic of the Paris Basin.

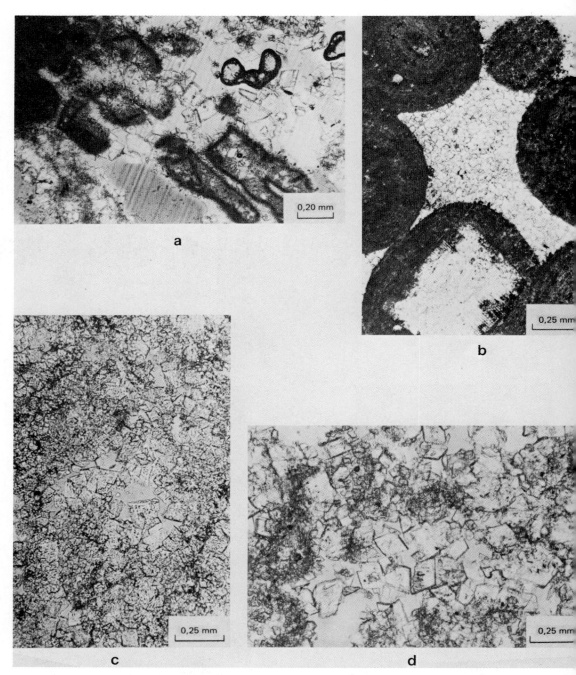

Figure 2-10 Relationship between porosity and the early diagenetic development of dolomite (thin-section photomicrographs in plane-polarized light). (*a*) Intraparticle porosity within a coral partially occluded by the development of dolomite crystals. Reef environment from the Mio-Pliocene of the Pacific. (*b*) Dolomite occluding original interparticle porosity in an oomicrosparite. Oölitic grainstone from the Jurassic of northeastern France. (*c*) Intercrystalline porosity developed by complete dolomitization of a reefal limestone. The dusty areas represent the original micritic margins of corals. Reef deposit from the Permian of Iran. (*d*) Example similar to (*c*) taken in a modern reef.

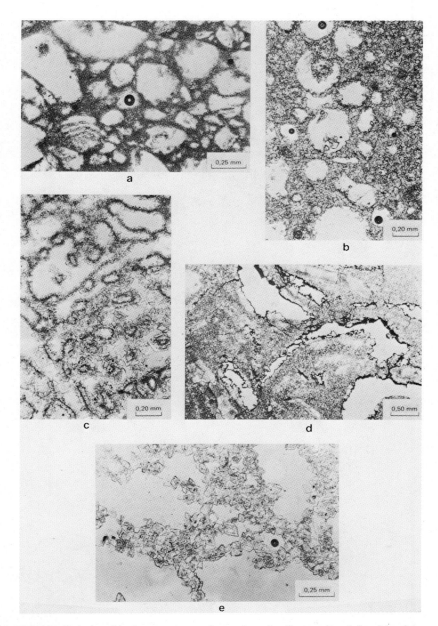

Figure 2-11 Relationship between porosity and early diagenesis—dolomitization and dissolution (thin-section photomicrographs in plane-polarized light). (*a*) Moldic porosity formed by dissolution of foraminiferal tests in a dolomicrosparite. Intertidal packstone from the Mio-Pliocene of the Pacific. (*b*) Moldic porosity formed by the solution of oöids in a dolomicrosparite. White patches containing scattered rhombs denote the former position of the oöids. Oölitic packstone to wackestone from the Permian of Iran (barrier or oölitic delta). (*c*) Moldic porosity resulting from dissolution of micritized particles. Dusty lines represent the remains of the particles. Bioclastic reef deposit from the Mio-Pliocene of the Pacific. (*d*) Moldic porosity resulting from the dissolution of bivalve shells in a dolomicrosparite. Intertidal deposit from the Gulf of Guinea; age unknown. (*d*) Vuggy porosity in a dolomicrosparite in association with intracrystalline porosity as a result of incomplete dissolution of the dolomite rhombs (see rhombic outline of crystal in the center of the photo). Mio-Pliocene of the Pacific.

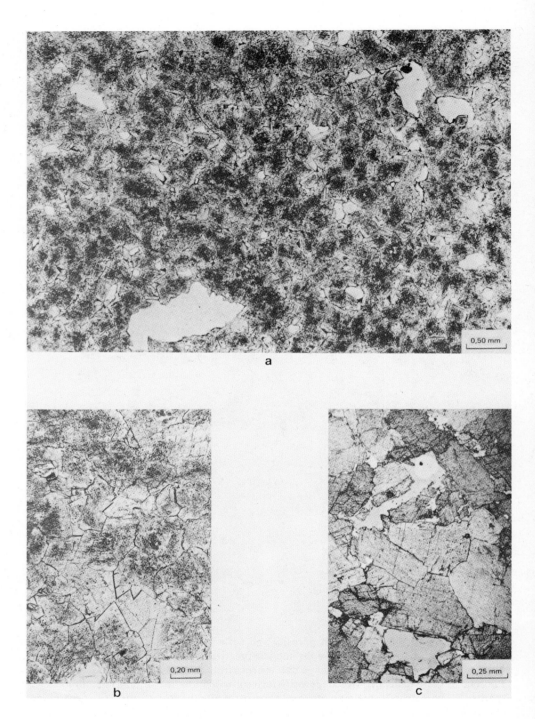

Figure 2-12 Relationship between porosity and dolomitization (thin-section photomicrographs in plane-polarized light). (*a*) Destruction of porosity by the growth of dolomite rhombs. Shelf deposit from the Upper Cretaceous of southwestern France. (*b*) Complete destruction of porosity by interlocking dolomite rhombs. High-energy deposit from the Jurassic of southeastern France. (*c*) Porosity formed by partial solution of dolomite rhombs. High-energy deposit from the Devonian of Belgium.

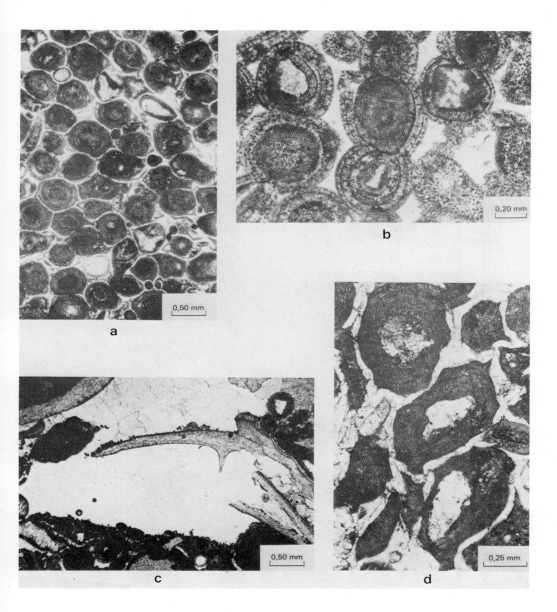

Figure 2-13 Relationship between porosity and burial diagenesis—compaction and cementation (thin-section photomicrographs in plane-polarized light). (*a*) Reduction in interparticle porosity as a result of compaction and minor pressure solution. Note local flattening of oöids with resultant fitted fabric and reduction in thickness of early grain-coating cement. Grainstone from the Jurassic of the Paris Basin. (*b*) Reduction in interparticle porosity caused by compaction and pressure solution at particle contacts. Note the flattening of the oöid cortex where two oöids are in contact, parting of the cortex from the nucleus, and the formation of stylolites at particle contacts. Oölitic shoal from the Jurassic of the Paris Basin. (*c*) Destruction of porosity by the precipitation of coarse calcite spar in a micropeloidal micrite containing bivalves. Biomicritic shelf deposit from the Jurassic of the Paris Basin. (*d*) Destruction of porosity by the precipitation of calcite in interparticle pores. The calcite may be liberated by pressure solution of adjacent oöids. Note the irregular and stretched shape of the oöids. Oosparite from the Jurassic of northeastern France.

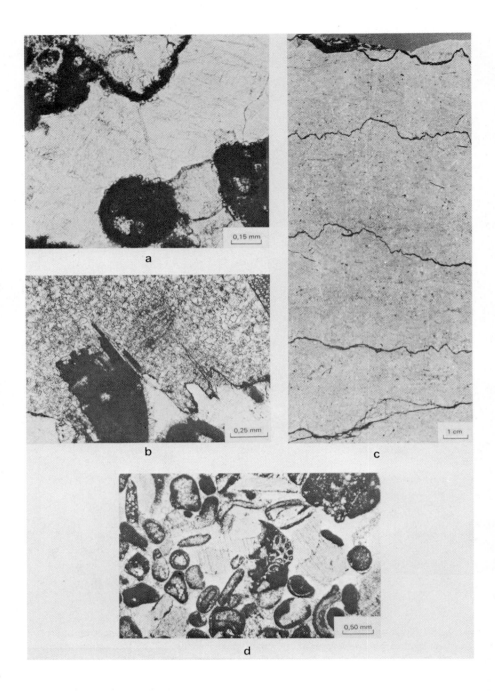

Figure 2-14 Relationship between porosity and burial diagenesis—pressure solution, cementation, and rim cement (photomicrographs of thin sections *a*, *b*, and *d*; polished core section *c*). (*a*) Interparticle pores filled by coarse sparry calcite cement. Oölitic shoal from the Middle Jurassic of the Paris Basin. (*b*) Irregular stylolitic contact overlain by dolomicrosparite. Jurassic grainstone from the Central Massif, France. (*c*) Stylolites developed in an inner-shelf packstone. Tertiary of Iraq. (*d*) Interparticle pores occluded by the development of calcite overgrowths on crinoidal debris. Coarse biosparite with oöids, crinoids, and sparse bryozoans and bivalves, high-energy shoal deposit, Jurassic of the Paris Basin.

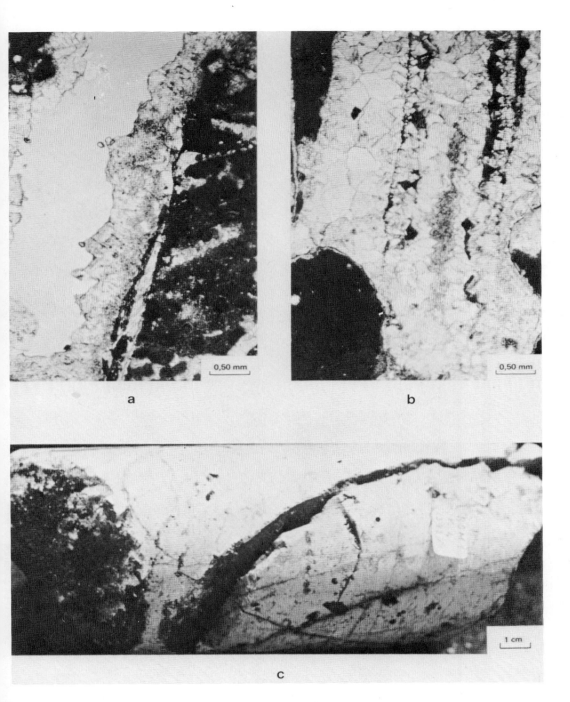

Figure 2-15 Relationship between porosity and burial diagensis—fracturing (thin sections impregnated with colored resin *a*, *b*, and core *c*). (*a*) Fracture bordered by calcite spar. The open fracture appears pale gray in color and imparts a high porosity on the coarse biomicrosparite that surrounds it. Shelf deposit from the Cretaceous of Italy. (*b*) Fractures healed by calcite. Fracture filling contains some argillaceous residue and shows the development of finer crystals in the center. The surrounding rock is a micrite. Shelf deposit from the Cretaceous of Italy. (*c*) Macroscopic view of an open fracture in a Cretaceous limestone from Italy.

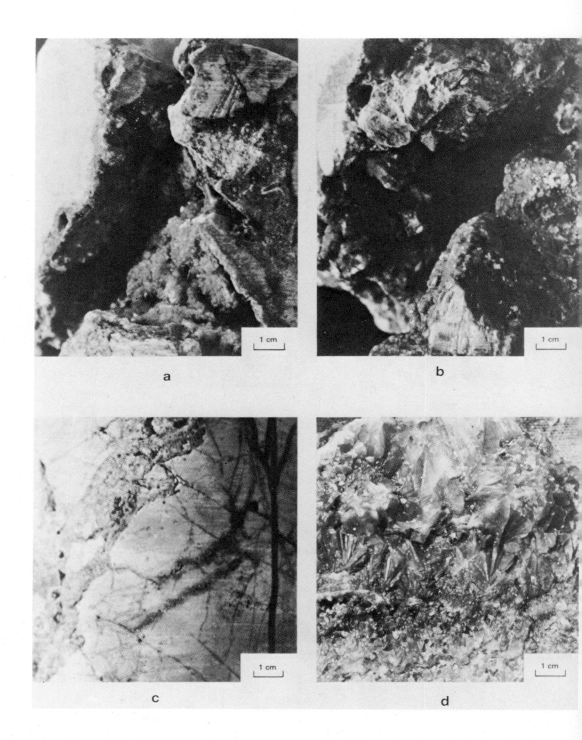

Figure 2-16 Relationship between porosity and late diagenesis—karst formation (photographs of cores). (*a*) A solution void of karstic origin partially filled by internal sediment. Cretaceous of Italy. (*b*) A similar void of the same origin from the Cretaceous of Italy. (*c*) Fracturing in Cretaceous limestone from Italy with subsequent karstic filling. (*d*) Cementation in a void of karstic origin from the Tertiary of Italy.

micropores and their connective network to be measured, and makes possible inference of diagenetic processes that have altered preexisting pores and connective networks. Textural analysis of the rock also allows more accurate pore determinations in three dimensions (Delfiner et al., 1972). More elaborate studies of certain aspects of the sample are used to solve specific problems; for example, a scanning-electron microscope can be used to examine fine pore networks, such as the interstitial or intraparticle porosity that occurs within chalks (Fig. 2-3*b*).

Well Logging

In the absence of cores, different well-logging techniques are used for continuous monitoring of the petrophysical characteristics of the reservoir. These physical parameters, whose variations are recorded with depth, depend on:

1   The mineralogy of the carbonate rock, whether limestone or dolostone.
2   The porosity and distribution of the pores.
3   The nature and percentage of fluids that occupy the pores.

The physical parameters as indicated by the log measurements are used to estimate porosity, hydrocarbon saturation, permeability, and lithology. The results obtained can be improved by laboratory calibration of the logs. Two of the most common kinds of logs are electric logs and radioactivity logs.

Electric Logs   These measure the spontaneous potential (SP) and electrical resistivity of strata in a well. The SP curve is basically a measurement of rock permeability and is related to the salinity of the fluid in the formation. The baseline of the SP curve is called the "shale line," and strata that register at this level include nonpermeable shaley units and units with fluid of the same salinity as the drilling mud. Readings vary around this baseline; negative (or deflections to the left of the shale line) for permeable strata containing salt water or oil, and positive readings or deflections to the right of the shale line for strata containing freshwater.

The resistivity of a rock unit is primarily a function of the amount of contained fluid and the electrical resistivity of that fluid. The amount of fluid present depends on the porosity, so that the resistivity curve is a function both of porosity and of the kind of fluid present. Oil, being a nonconductor, has high resistivity, while a unit containing salt water has low resistivity.

Radioactivity Logs   The gamma-ray and neutron log, are the main kinds of radioactivity logs. The gamma-ray log records the natural radioactivity of the rock. Radioactivity is generally high for shales and low for limestones. The neutron log measures radioactivity generated by the bombardment of strata with neutrons and is basically a function of porosity. For a more detailed discussion of well-logging techniques see LeRoy, LeRoy, and Raese (1977) and Asquith (1979).

Productivity Measurements

Measurements of productivity are most precise when determined directly in the borehole. Their significance depends on how they were done and on the quality of the background information, that is knowledge of lithologies present, sedimentary framework (the geometry and extent of favorable reservoir horizons), initial flow-rate information (need for perforation, fracturing, or acidification to improve flow if the cement is calcite), or knowledge of the nature of the pores and their distribution.

The methods for measuring productivity are varied. Some are quantitative while others are semiquantitative (see Chilingar et al., 1972; LeRoy, LeRoy, and Raese 1977). The choice of methods depends on the type of problems posed.

KINDS OF PORES IN CARBONATE RESERVOIRS

Classification of Porosity

No one classification of pore types is completely satisfactory because of the complexity of carbonate rocks and the heterogeneity of carbonate reservoir horizons. Variations exist on all scales from differences between sedimentary bodies to heterogeneity of pore connections, pore size and shape, to variations in micropore characteristics. The scale at which the pores are studies depends on the problem under investigation.

Various classifications of porosity are given in the Appendix to Chapter 2. Choquette and Pray (1970) offer one of the most comprehensive classifications, which is used as a basis for this section. The genesis of pores is dealt with in a later section.

Two basic kinds of pores are recognized in carbonate rocks: (1) primary pores (pores that form in the depositional environment and are responsible for primary porosity) and (2) secondary pores (pores that form in a sediment or rock after final deposition and constitute secondary porosity). The following classification is used here:

1   **Primary porosity**
    (a)   Framework
              Intraskeletal
              Interskeletal
    (b)   Intraparticle
    (c)   Interparticle
              Between sand and pebble-sized particles (interskeletal, interoölitic, inter-peloidal, interrock fragment, interintraclastic, shelter porosity, etc.)
              Between silt- or clay-sized particles (intramicrite-matrix porosity of some workers, intercoccolith-chalk porosity of some workers, intralime mud, etc.)
    (d)   Fenestral
2   **Secondary porosity**
    (a)   Intercrystalline
    (b)   Moldic
    (c)   Vug, channel, and cave
    (d)   Fracture, breccia
    (e)   "Chalky" or weathering porosity

Primary Porosity

Framework Porosity (Fig. 2-2c)

These are the inter- and intraskeletal pore spaces that occur between and within organisms in a bound carbonate framework (boundstones). Such frameworks include coral reefs, oyster banks, rudistid banks, and buildups of algae, bryozoans, and sponges. The pore spaces in these buildups (framework porosity) are commonly partially filled with fibrous or cryptocrystalline cements and internal sediment.

Intraparticle Porosity (Figs 2-2c and 2-3d)

These are pore spaces that occur within particles. In skeletal fragments they are usually formed with the disappearance of the soft organic parts from the carbonate skeleton, for example, removal of the polyps from the coral skeleton or organic tissue from foraminifera.

Interparticle Porosity (Figs. 2-2a, 2-2c, and 2-2d)

This kind of porosity corresponds to the spaces or voids between particles and is that normally found in siliciclastic sandstones. In carbonate sediments this kind of porosity varies with the type of particles present and is modified by later compaction or introduction of internal sediment and cement. The shapes of the pores vary greatly and are homogeneous only when the enclosing particles are of uniform shape and size; for example, well-sorted oöids. Interparticle porosity not only occurs between large sand- and pebble-sized grains, but also in lime muds and chalks (Fig. 2-3b). Certain kinds of pores can be considered to be interparticle.

Shelter Porosity (Fig. 2-3c)   This kind of porosity (also called umbrella voids) is found below large particles, such as skeletal fragments or intraclasts. The large particles act as "umbrellas," protecting the pore space beneath them from the downward movement of fine material. This kind of porosity is common in calcarenites which contain large mollusc shells, large foraminifera, or corals. The pore spaces beneath the large particles may be partially filled by internal sediment that forms geopetal surfaces. These flat surfaces allow the orientation of the rock to be reconstructed for the time at which the internal sediment was emplaced.

Breccia Porosity   Primary breccia porosity occurs between the fragments of a depositional carbonate breccia. The fragments forming the breccia may be intraclasts (interintraclastic porosity) from within the depositional environment or rock fragments carried in from elsewhere. This kind of breccia porosity may be reduced by the presence of a matrix or cement.

Fenestral Porosity (Fig. 2-3a)

These kinds of pores are generally elongate to equant and are primary gaps in the rock framework larger than the grain-supported interstices. They may be arranged in layers throughout the rock or occur in an irregular manner. Fenestral fabrics, also known as bird's eye fabric, are interpreted as the voids formed in algal mat sediments through desiccation, rotting away of algal tissue, and production of gas bubbles.

Secondary Porosity

Intercrystalline Porosity

This kind of porosity is the result of spaces that occur between crystals and is most important in dolomitic rocks (Figs. 2-10c and 2-10d). Much of the dolomite is of replacement origin, and the development of intercrystalline porosity results from the

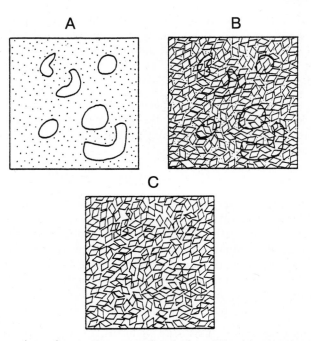

Figure 2-17 Formation of secondary porosity during dolomitization by the leaching of intercrystalline CaCO₃ remnants (from Friedman, 1980, Fig. 16. p. 604).

growth of randomly oriented rhombs coupled with dissolution of interstitial Ca₃CO which has not been replaced (Fig. 2-17).

Moldic Porosity (Figs. 2-7a, 2-11a, 2-11b, 2-11c, and 2-11d)

Moldic porosity is formed by the selective removal of primary constituents from a sediment or rock, such as ooids, bivalve shells, and evaporite minerals. Molds are distinguished by their shape, size, wall features, or other relict features that indicate the former presence of a particular kind of particle or crystal.

Vug and Channel Porosity (Figs. 2-5d, 2-6e, 2-7b, 2-8a, 2-8b, 2-9b, 2-11e, 2-16a, 2-16b, and 2-16d)

This kind of porosity results from dissolution and may bear no relationship to the initial rock texture. The voids formed are of irregular shape and size and may or may not have interconnections. Many vugs may be solution-enlarged molds where evidence of the precursor has been destroyed by enlargement. Cave porosity is a form of vug porosity and is distinguished by the large size of the pores.

Channel porosity is distinguished from vug porosity by the geometry of the pores. Channel porosity consists of tabular to flat channels which augment any other porosity present in the rock and allow drainage of the strata, thus enhancing permeability.

Fracture and Breccia Porosity (Figs. 2-15a, 2-15b, 2-15c, and 2-16c)

This kind of porosity is more common in dolostones that fracture readily under tectonic stress; limestones tend to yield by flow and pressure solution. In recent years it has been shown that microfissure networks (openings averaging 20 μm in diameter) develop in the same way as the fracture networks and are responsible for draining a considerable volume of rock. Fracture porosity is well developed in ancient buried rock (a late-stage diagenetic event), but also appears very quickly in younger carbonate rocks (Plio-Pleistocene) with only a shallow cover. Fracture porosity grades into secondary breccia porosity with increasing displacement, disruption, and collapse of strata.

Fracture porosity can exist on its own in compacted rocks and can be the single factor responsible for reservoir existence. It can also coexist with other kinds of porosity and provide essential permeability. The following variations can exist:

1   Pure fracture porosity.
2   Fracture porosity associated with some other kinds of pores.
3   Fractured pores.

No adequate classification exists yet for complex fracture or fissure networks. The following variations in fracture patterns are known to occur:

1   Different fracture intensities and lengths.
2   Variations in fracture width.
3   Existence of several fracture networks.
4   Continuity of fractures across beds.
5   Links with other kinds of pores.
6   Links with stylolites.
7   Filling by secondary cements (completely filled fractures are also known as "healed fractures").

Fractures can be studied in many ways. These include statistical analysis, lithological and petrological examination, and geophysical tracing. Fissures and fractures play an essential role in reservoir drainage.

"Chalky" or Weathering Porosity

Carbonate rocks and sediments commonly weather when subaerially exposed to form a friable, "chalky" sediment. This is common in reef rocks and skeletal sands in modern tropical and subtropical settings. Some of the world's largest oil and gas reservoirs occur in subsurface carbonates which have been "chalkified" during their geological history.

RELATIONSHIPS BETWEEN DEPOSITIONAL ENVIRONMENTS AND RESERVOIRS (Figs. 2-2, 2-3, and 2-18)

Studies of porous networks show that the petrophysical characteristics of a rock depend on conditions of sedimentation and on diagenetic alterations. Loss of porosity can occur early in diagenetic history as the result of changing conditions in the environment of deposition or at a late stage when influenced by burial. An initial pore network in a carbonate sediment essentially depends on high kinetic energy in the environment of deposition (Fig. 2-2*a*) as well as on the growth form of the organisms.

High kinetic energy is directly responsible for the coarse interparticle porosity found in the shoaling zone. Fine material is carried out of this environment into the subtidal zone, leaving a lag concentrate of sand-sized carbonate particles with interparticle porosity. However, in consistent low-energy environments the accumulation of microorganisms or their broken shells can result in a very fine interparticle porosity.

The growth form of organisms can contribute markedly to primary porosity. Framework porosity is strongly controlled by the growth forms of the frame-building organisms, such as corals in particular. However, this kind of porosity is usually altered greatly by diagenesis (Figs. 2-6c and 2-8c) and is rarely preserved in its original state.

Fenestral porosity is usually attributed to algal mat development and consists of the void spaces left by gas bubbles or by the decay of organic matter after lithification of the surrounding sediments.

Intraparticle porosity is extremely important in some deposits such as rudistid reefs, concentrations of foraminifera (Fig. 2-3d), or bivalves.

Primary porosity, such as interparticle porosity, prevails in specific depositional settings. In coarsening-upward sequences caused by increasing energy the most favorable site for a reservoir with primary porosity would be at the top of the sequence. In fining-upward sequences the best potential reservoir rocks would be in the highest-energy deposits at the base of the sequence. The following basic sequences can be recognized in the inner-shelf supratidal and intertidal environments:

1 Fining-upward tidal channel and lagoonal sequences extend from the subtidal to the supratidal environment in the inner shelf. The grainstones and packstones at the base of the sequence may possess sufficient initial porosity as a result of high-energy deposition. However, in the supratidal deposits at the top of the sequence, which were not originally porous, freshwater leaching on the subaerial exposure may improve reservoir potential by creating secondary porosity.

2 Inversely, coarsening-upward sequences that progress from the supratidal to subtidal environment have the best reservoir potential, that is, primary interparticle porosity, at their top.

In the outer shelf, slope, and deep ocean basin environments the following sequences can be distinguished:

1 Accretionary or regressive sequences (open ocean to barrier or shoestring sand) are coarsening-upward sequences produced by the lateral migration of environments into the basin. The shallowest deposits, which are also probably in part emergent, are also the highest-energy sediments and thus form the most favorable reservoir horizons with primary interparticle porosity.

2 Deepening or transgressive sequences are the opposite. Energy decreases upward at any one point, and a fining-upward sequence is produced. The barrier is displaced shoreward. A barrier reef buildup presents the best potential reservoir because it forms in the highest-energy environment in a zone of favorable diagenetic alteration, notably dolomitization.

3 On the slope and in the basin proximal and distal turbidite sequences occur and also include potential reservoir horizons at their base.

RELATIONSHIPS BETWEEN DIAGENESIS AND RESERVOIR DEVELOPMENT

Primary porosity forms as a result of processes of sedimentation (Fig. 2-18). Pore destruction and the development of secondary porosity occur with changing diagenetic conditions at various stages in the history of a deposit.

Relationship Between Diagenetic Stages and Reservoir Development (Table 2-1)

Changes in a sediment as the result of interaction with near-surface waters may modify or eliminate primary porosity. These waters have different patterns of distribution in different settings (Fig. 2-19) and can be divided into a number of zones:

1 Vadose zone, the zone above the groundwater table.
2 Freshwater phreatic zone below the groundwater table.
3 Marine phreatic zone.

The vadose zone is characterized by intermittent flow of freshwater. Through dissolution it may become an excellent site for pore formation; only minor cementation occurs at particle contacts. Vadose cement is usually calcite. Climatic influence can considerably affect the relationship between solution and cementation in the vadose zone.

Being below the water table, the freshwater phreatic zone is characterized by permanent freshwater flow (Fig. 2-19). Compared to the vadose zone, cementation in the phreatic zone can be significant. Aragonite is not preserved in the freshwater phreatic zone, where waters are saturated with $CaCO_3$ and have a low Mg^{2+}/Ca^{2+} ratio. Drusy calcite spar is precipitated within pore space and progressively seals both intra- and interparticle porosity (Figs. 2-8c and 2-9b). Any earlier aragonitic cement as well as primary aragonitic particles are replaced by a calcite spar (Figs. 2-6d and 2-8c), either by solution and void filling or by neomorphic inversion.

The marine phreatic zone results from invasion of the sediment by seawater. The influence of marine water on sediment diagenesis is not fully understood, but the zone of mixing between the marine and freshwater is an important interface along which cementation may occur.

The importance of these three near-surface water zones depends heavily on climate which influences movement of water and evaporation, on the morphology of the emergent area, and on the drainage conditions (porosity, permeability, or dimension and geometry of the sediment bodies). Under certain conditions the freshwater vadose zone occurs directly above a marine phreatic zone (Fig. 2-19c). Mixing of waters does occur, and the boundaries between the zones change with time. These interactions of the water bodies have important consequences on the distribution of zones where solution, cementation, and dolomitization occur. Despite the importance of diagenesis in these near-surface water zones to the formation and destruction of potential petroleum reservoirs, the actual sequence of events due to the interaction of these water bodies is poorly understood.

When diagenetic alterations take place after burial below the zone of surface waters

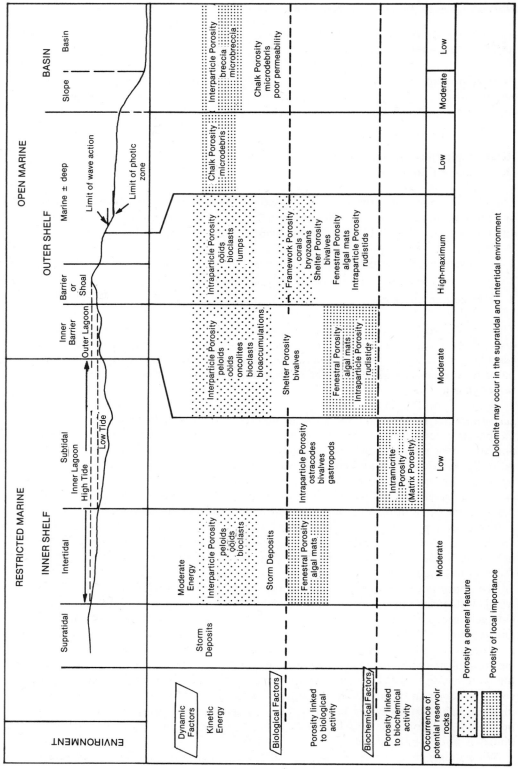

Figure 2-18 Distribution of primary porosity related to depositional environment and dynamic, biological, and biochemical factors.

66

Table 2-1 Relationship between the Stages of Diagenesis and Creation or Destruction of Porosity

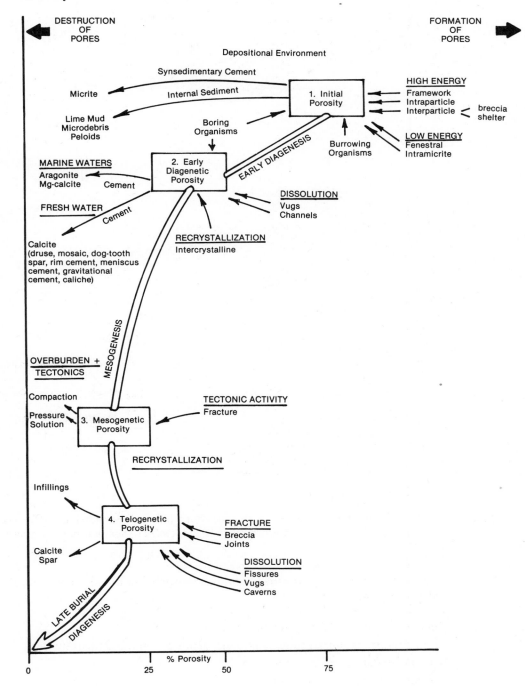

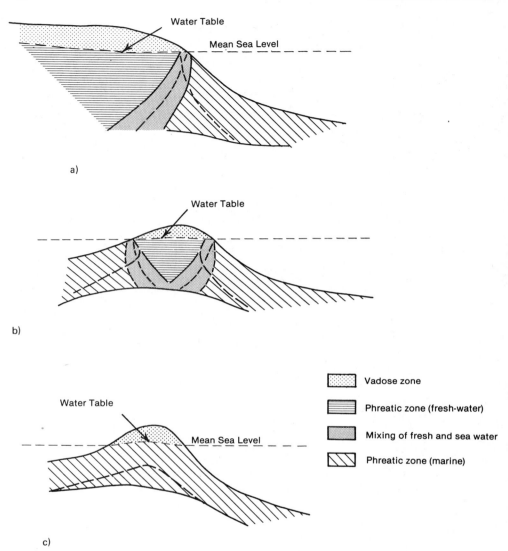

Figure 2-19 Distribution of vadose and phreatic zones. (*a*) Near the shoreline. (*b*) In a carbonate island, reef, or shoal. (*c*) In a carbonate island, reef, or shoal where the freshwater phreatic zone is absent.

(mesogenesis) the time taken for the alterations to occur can be extremely long. The diagenetic events are controlled by:

1 Changes in the sediments prior to burial.
2 Burial conditions (burial time, maximum depth of burial, temperature).
3 Tectonic activity (pressure, stress).
4 The nature of the connate waters.

Finally, the rock may again be brought to the surface where late subaerial diagenesis occurs (telogenesis).

Early Diagenetic Changes (*Eogenesis*)
(*Fig. 2-20 and Table 2-2*)

When conditions are favorable for diagenesis initial reservoir characteristics can be modified quickly before or after lithification. Processes that modify porosity prior to lithification include bioturbation, burrowing, algal and bacterial processes, and dissolution of organic matter. All these processes have important influence on the development of interconnected pore networks, whereas cementation by crypto-crystalline cement destroys some of the interparticle porosity in high-energy deposits (Friedman and Sanders, 1978, p. 175). After lithification the development of a pore network depends on the activity of organisms (Fig. 2-4*b*) and microorganisms, physical and chemical factors, in an environment either comparable to the one in which deposition took place or different from that of initial deposition. Whichever situation occurs, the nature of the original sediment and the original pore network has a large influence on the diagenetic evolution.

The diagenetic alterations that affect the lithified and unlithified sediments change with time and with depth of burial. At the sediment/water interface biological and physical agents predominate.

1 Organisms have a direct effect on porosity (Fig. 2-4). They create a pore network inside soft sediments (burrowers, e.g., worms, crabs) or in lithified sediment (rock-boring molluscs). Their activity is confined to near the surface of the sediment.

2 Microorganisms (algae and bacteria) have an indirect effect. They attack the surface of particles, which they severely corrode or micritize. They create conditions that are favorable to later dissolution or protect the particle against epitaxial overgrowth. This activity would appear to be more prolific near the surface of the sediment, but may extend down into already buried sediment. Together with kinetic energy, the activity of microorganisms contributes to the fragmentation and destruction of skeletal particles.

The dynamic action of water continues to affect sediment after deposition. Existing deposits may be reworked to form intertidal breccias and flat-pebble conglomerates; fine lime mud may be trapped within or infiltrate preexisting porous deposits in high-energy zones (Fig. 2-5). The latter process can lead to the filling, cementation, and complete obliteration of porosity in potential grainstone and boundstone reservoirs; micropeloidal micrite is common in bioclastic sands, whereas micrite and very fine skeletal debris are common in the voids of boundstone frameworks. Lime mud may also occlude burrow networks, borings, and voids created by dissolution.

Disintegration of organic matter in calcareous skeletons produces primary intraparticle porosity. In algal-mat deposits the decay of organic matter contributes to the formation of fenestral porosity when the host sediment is already partially lithified and can resist collapse.

Desiccation is an important superficial process that can generate a shrinkage-fracture system and may cause brecciation.

Physical and chemical processes involved in the alteration of sediments are not necessarily limited to the sediment surface. Changes are mainly induced by interaction with circulating waters.

Marine waters can cause cementation by precipitation of fibrous aragonite and fibrous or cryptocrystalline high-Mg calcite (Friedman, Amiel, and Schneidermann, 1975). Generally these waters have a weak diagenetic effect, except in reef buildups where marine cementation is an important process.

	Material Involved	Process	Result — Pore Formation	Result — Pore Destruction	Environment
PHYSICAL FACTORS	Surface Sediment	Desiccation	Porosity in shrinkage cracks and breccia	Internal sedimentation —micritic peloids, micrite	Supratidal Intertidal
BIOLOGICAL	Soft Lime Mud Algal Mats Hardgrounds	Burrowers Gas Bubbles Borers	Vuggy network porosity Fenestral porosity Favors dissolution between bioclasts ← algal and bacterial micritization		intertidal
BIOCHEMICAL	Organic Matter	Bacteria	Interparticle porosity Intracoralline porosity Fenestral porosity Micropore porosity (within algae)		Intertidal Subtidal
CHEMICAL	Evaporites	Meteoric Water	Dissolution of sulfates and halite. → fissure porosity → vuggy porosity → moldic porosity → solution-collapse breccia porosity	Cementation/Infilling " " " "	Supratidal
	Aragonite	Marine Water Marine/ Fresh-water	Dissolution intracoralline porosity moldic porosity	Cementation fibrous aragonite Transformation aragonite-calcite	Intertidal
	Calcite	Meteoric Water	Dissolution of outer surface: → "chalky" porosity Pore opening dissolution: → conduit porosity Massive dissolution of calcite: → vuggy porosity Massive dissolution of dolomite: → vuggy porosity Recrystallization: → microcrystalline porosity → intercrystalline porosity	Micritic internal sedimentation Formation of caliche Equigranular mosaic (within organisms) Drusy mosaic	Supratidal Emergence
	Dolomite	Hypersaline Waters Marine/ Fresh-water Mixing	calcite → dolomite → intercrystalline porosity Transformation → intercrystalline porosity Dissolution → vuggy porosity → pseudomorph porosity	Early dolomitization Cementation of lime mud Recrystallization of dolomite Crystal growth Interlocking dolomite mosaic	Intertidal to Supratidal

Freshwater can have profound diagenetic effects and, depending on mineral composition, different processes occur:

1 Aragonite is unstable in freshwater and disappears early during freshwater diagenesis (Friedman, 1964; Bathurst, 1966) (Fig. 2-6). Dissolution of aragonite is spectacular and may result in complete removal of skeletal material (moldic porosity). Replacement of aragonite by calcite is a complex solution-precipitation reaction on a microscale and leads to a decrease in particle microporosity. Locally fibrous aragonitic cementation can occur in the zone of mixing between marine and freshwater, while in the vadose zone growth of an oriented fabric such as vadose pisolites (Dunham, 1969) may occur. The disappearance or transformation of aragonite is an important feature in establishing a new physiochemical equilibrium within the rock and in extensively altering the pore network.

2 Magnesium calcite can act as a cryptocrystalline cement. It is also an important site for mineral transformation. High-Mg calcite transforms to low-Mg calcite with liberation of Mg^{2+} ions.

3 Evaporites are very soluble and are easy minerals to dissolve (Fig. 2-7). In sediments under the influence of vadose or saturated waters evaporites can fill the pores. They are precipitated in the existing pore spaces filling molds, fissures, intercrystalline spaces, and so on, and even displace the sediment as a result of nodule formation. Dissolution of evaportites reopens these pore spaces or forms a solution breccia.

4 Calcite is also sensitive to the influence of freshwater. The solubility of calcite increases with the amount of CO_2 dissolved in the water. Continental waters of meteoric origin tend to have a low pH because of the presence of H+ ions and readily dissolve calcite.

To dissolve previously stable carbonate particles all that is required is the emergence of lithified marine sediment into a zone of freshwater. When these waters are no longer in contact with the atmosphere the pH increases and the waters become alkaline. Under conditions of high alkalinity calcite and aragonite can precipitate from waters, even from waters with low concentrations.

Dissolution is an important factor in producing variations within reservoir rocks (Fig. 2–8). Micrite can be dissolved with accompanying enlargement of existing pores, development of a connective network between pores, and enlargement of pore throats. The outer surfaces of crystals may dissolve, creating a friable texture and a porosity like that found in "chalkified" limestones. Dissolution of calcite remnants from dolostones creates intercrystalline porosity (Fig. 2-17).

Cementation has varied forms (Fig. 2-9). In the intertidal environment cementation is restricted to a coating around particles. In the supratidal zone the initial cement may be a disordered layer of crystals coating a void (dogtooth spar) followed by precipitation of a drusy mosaic that fills interparticle void spaces. Calcite also is deposited as equant crystals filling the chambers of organisms. This kind of cementation is important in the obliteration of any pore network.

Dolomite can form as a replacement under conditions that are still poorly understood. Two approaches to the problem are:

1 A theoretical concept which suggests that a solid state transformation from $CaCO_3$ to dolomite would result in a solid volume reduction of 5.8% from aragonite and 12.9% from calcite.

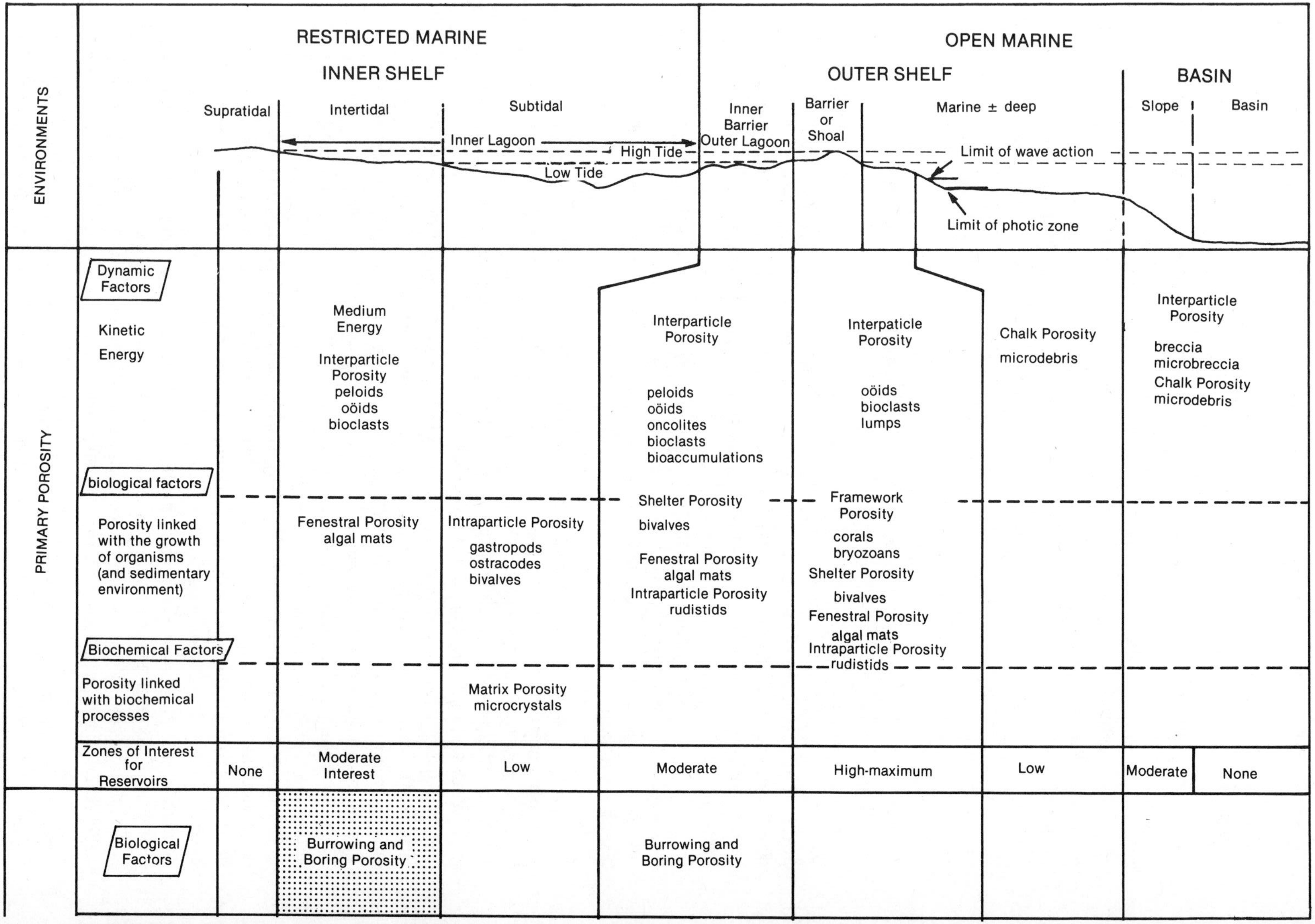

ENVIRONMENTS

	RESTRICTED MARINE — INNER SHELF				OPEN MARINE — OUTER SHELF		BASIN	
	Supratidal	Intertidal	Subtidal (Inner Lagoon)	Inner Barrier / Outer Lagoon	Barrier or Shoal	Marine ± deep	Slope	Basin

Limit of wave action — High Tide — Low Tide — Limit of photic zone

PRIMARY POROSITY

	Supratidal	Intertidal	Subtidal	Inner Barrier / Outer Lagoon	Barrier or Shoal	Marine ± deep	Slope	Basin
Dynamic Factors — Kinetic Energy		Medium Energy; Interparticle Porosity: peloids, oöids, bioclasts		Interparticle Porosity: peloids, oöids, oncolites, bioclasts, bioaccumulations	Interparticle Porosity: oöids, bioclasts, lumps	Chalk Porosity microdebris	Interparticle Porosity: breccia, microbreccia, Chalk Porosity microdebris	
biological factors — Porosity linked with the growth of organisms (and sedimentary environment)		Fenestral Porosity algal mats	Intraparticle Porosity: gastropods, ostracodes, bivalves	Shelter Porosity bivalves; Fenestral Porosity algal mats; Intraparticle Porosity rudistids	Framework Porosity: corals, bryozoans; Shelter Porosity: bivalves; Fenestral Porosity: algal mats; Intraparticle Porosity: rudistids			
Biochemical Factors — Porosity linked with biochemical processes			Matrix Porosity microcrystals					
Zones of Interest for Reservoirs	None	Moderate Interest	Low	Moderate	High-maximum	Low	Moderate	None
Biological Factors		Burrowing and Boring Porosity		Burrowing and Boring Porosity				

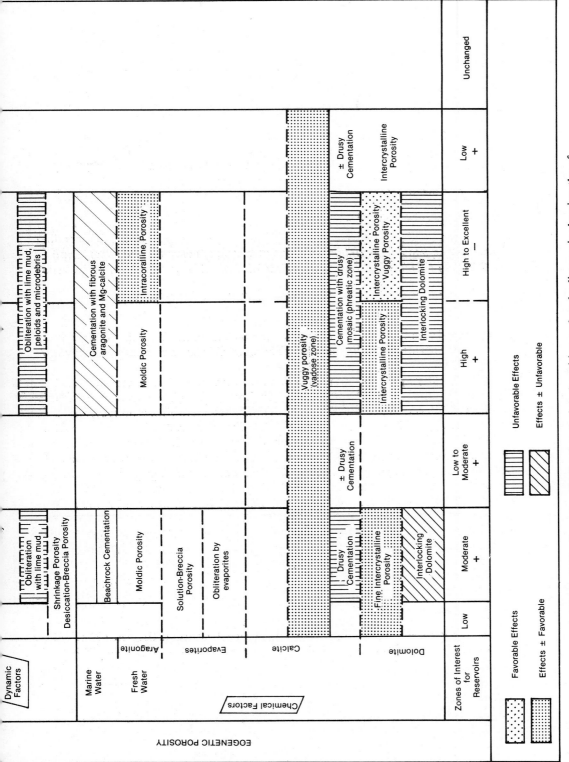

Figure 2-20 Porosity distribution during deposition and early diagenesis showing the effects of dynamic, biological, biochemical, and chemical factors.

2   A practical concept which takes into account the inference of most sedimentologists that the transformation takes place through an intermediate phase of dissolution. The replacement of aragonite or calcite by dolomite creates a solubility contrast in freshwater between the more soluble $CaCO_3$ and the less soluble dolomite. Freshwater dissolves the $CaCO_3$ that has not yet been replaced and thus creates intercrystalline porosity between the dolomite rhombs.

Many hypotheses deal with the origin of dolomite; some of these hypotheses are even contradictory. For further details on dolomite and dolomitization see Friedman and Sanders (1967) and Zenger, Dunham, and Ethington (1980).

Stages of Burial Diagenesis (Mesogenesis)
(Table 2-3, Fig. 2-21)

Burial begins when the sediment is no longer under the influence of surficial waters and ends when the sediment is again brought to the surface. During burial the sediment is under the influence of different factors depending on:

1   Depth, temperature, water flow, and burial time.
2   Tectonic activity and its intensity, duration, and nature.

These factors tend to modify the initial porosity, either obliterating it completely or increasing it by developing a network of fractures.

Sediment compaction occurs after deposition and continues with time (Fig. 2-13). When the sediment is still soft, plastic deformation is common. In lithified sediments the particles tend to be rearranged and the interparticle pores may progressively decrease in size. Fragile fragments such as bivalve shells may break. When a rigid framework is established the particles begin to deform and flatten. Pressure solution forms microstylolites at particle contacts, and the pore volume is further decreased. The calcite that is thus dissolved contributes to cementation and filling of pore space.

The effect of pressure solution is initially seen at particle contacts. After the pore volume has been greatly reduced stresses are transmitted throughout the rock rather than just at particle contacts. Stylolites develop (Fig. 2-14) together with fractures from compression (closed) and tension (open). The orientation of both the fractures and the stylolites depends on the directions of the acting stresses from burial and tectonic movement. Calcite liberated by pressure solution contributes to the cement that fills pores and heals fractures.

Under the influence of increasing temperatures and pressures carbonates stable at an earlier diagenetic stage tend to be changed. These changes are of a chemical nature (transformation, loss or gain of magnesium) or of a crystallographic nature (epitaxial overgrowth, recrystallization, micritization) and may result in a loss of porosity from crystal growth and rearrangement. Occasionally the opposite occurs; accompanying dolomitization porosity may develop. Previous diagenetic textures are altered and their relative importance and even their existence as factors involved in the formation of the pore network are no longer evident.

Epigenetic dolomitization is related to the presence of faults or fractures (a source of fluids) and may contribute to the improvement of porosity in a reservoir at right angles to the stylolites and fissures.

Table 2-3 The Effects of Burial Diagenesis and Late Subaerial Exposure on the Porosity in Reservoirs

| | Process | Effects on Reservoir Strata | Kinds of Pores/Effects Upon Pores | |
			Pore Destruction	Pore Formation
MESOGENESIS	growth	epitaxy (rim cement)	decrease, disappearance of pores	
	compaction compression	plastic deformation	decrease in porosity	
		rearrangement of grains	decrease in porosity	
		broken grains	decrease in porosity	
	pressure solution	at grain contracts } deformation stylolitic contacts	decrease in pore openings	
			decrease in pore openings	
		stylolitization	decrease in porosity by cementation as a result of the liberation of calcite/dolomite	stylolite porosity
	tectonic movement	compressional cracks	decrease in porosity	
		tensional cracks		fracture porosity
	cementation	coarse calcite	obliteration of porosity	
	recrystallization			intercrystal porosity
	dolomitization	localized along fractures and faults		intercrystal porosity
TELOGENESIS	fracturing (unloading)	joint formation	infilling with sediment and spar (healed fractures)	fracture network porosity
	karstic dissolution	fissures, vugs, caves	sparry cementation cave filling	karstic and vuggy porosity
	recrystallization			intercrystal porosity
	dedolomitization	coarse dedolomitization in patches	patchy cementation	

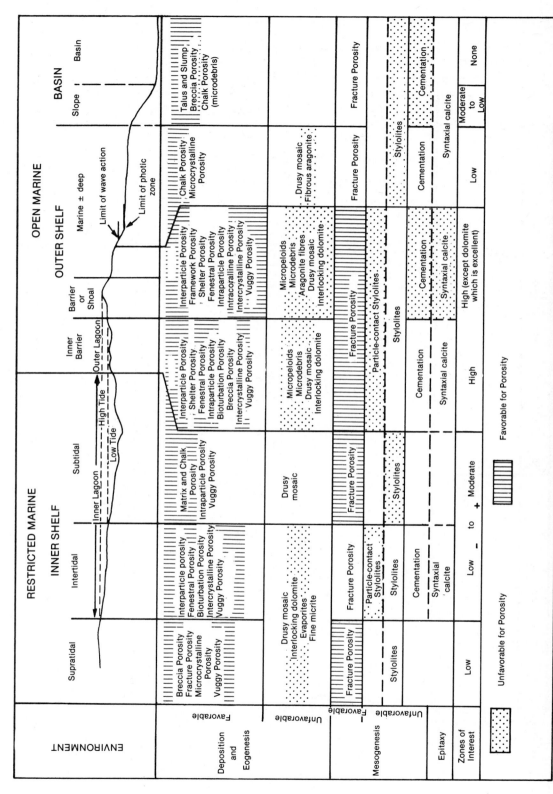

Figure 2.21 Diagenesis and pore formation during burial

76

Far from being restricted to burial diagenesis, the development of rim cements may be an early diagenetic feature. Epitaxial overgrowths around echinoderm debris can occur at any time and is related to the composition and nature of the pore fluids in the sediment (Fig. 2-14*d*).

Burial tends to obliterate the pore network in carbonate sediments, particularly in zones where the initial porosity was favorable for reservoir formation (Fig. 2-21).

Late Subaerial Alteration (Telogenesis)
(Table 2-3, Fig. 2-16)

Following burial, lithification, and compaction, sedimentary rocks may be reexposed. Many processes are important in modifying carbonate rocks at this stage.

1 Decompression of the rocks as they adjust to removal of overburden can cause extensive fracturing and joint formation.
2 Interaction with waters percolating through fractures produced during burial can result in intense corrosion and dissolution of the rocks. This can lead to the formation of caves, large vugs, and fissures (karstic porosity). Infillings of collapse breccia or massive aragonitic crusts can modify this kind of porosity, where active freshwaters or marine waters flow as underground rivers or are ponded in caves. Mineralogical changes such as dolomitization or dedolomitization can also occur.
3 Pedogenesis as a result of climate may develop a very large pore network.

Pore Formation and Destruction as a Result of Diagenesis (Table 2-1)

Pore Formation

Inherited, Fossilized, and Stabilized Porosity   This simplest kind of porosity is that inherited from originally porous particles, such as those found in foraminifera or corals. This primary porosity exists before deposition. A second kind of primary porosity is that found between particles and framework builders. Tufas formed as precipitated crusts on growing frameworks can help to preserve original porosity almost completely.

Primary porosity is a stabilized porosity that corresponds to the optimum grouping and packing of particles according to their shape and size. It may be reduced by compaction.

Primary porosity is mostly linked to high-energy environments, where grainstones are deposited and where silt- and clay-sized particles can be removed. However, some carbonates that originate in low-energy environments, such as chalks, have a high microporosity.

Porosity Formed by the Removal of Organic Matter   The rotting away and disintegration of organic matter can leave gaps within a carbonate sediment, such as in algal mats. These voids, known as fenestral or bird's eye pores, develop in the supratidal and intertidal environments. Voids produced by escaping gas bubbles formed by bacterial decay of organic matter explain porosity that cuts across layering in this kind of laminite.

Porosity Created by Organisms   Organisms influence porosity, either directly or indirectly. They form burrow networks or vuggy porosity as they move through a sediment. They also modify existing porosity by mixing coarse and fine sediments and by fragmenting existing particles.

Porosity of Mechanical Origin   Porosity in carbonate sediments or rocks during lithification may be caused by:

1   Dissolution of strata (solution breccia).
2   Early compaction and breakage of particles.
3   The formation of joints or faults.
4   The formation of fracture or fault breccias.

Solution Porosity   Porosity can be generated by dissolution in rocks that are not in equilibrium with their interstitial waters. Changes in equilibrium can be brought about by changes in pH, temperature, concentration of ions in solution, and other factors. Dissolution produces pipes, channels, and regular voids where individual particles or crystals have been dissolved; examples include oöids, shells, anhydrite, gypsum or halite crystals, and dolomite rhombs. In places irregular voids are created where the entire rock is dissolved, as in karstic dissolution.

Pore Destruction

Processes that destroy porosity are more numerous and more effective than those that create it.

Carbonate Cements   Carbonate cement as pore fill may have diverse origins. Needle-shaped aragonitic cement and cryptocrystalline high-Mg calcite cements are frequently precipitated by marine waters, usually either in reefs or in beachrock (Friedman, 1975).

    The most common kind of cement is low-Mg calcite. It occurs as a drusy mosaic of calcite crystals which apparently form in the phreatic zone under the influence of freshwater (Bathurst, 1975), as dogtooth spar which also forms under the influence of freshwater but is not important in changing reservoir quality, and as a meniscus cement which is typically found in the vadose zone (Dunham, 1971). Other kinds of calcite cement with diverse origins occur and play an important part in pore destruction.

Changes in Mineral Composition   Changes from aragonite to calcite and from calcite or aragonite to dolomite are common. They can lead to a reduction or disappearance of the pore network. Where aragonite changes to calcite, microporosity and accompanying permeability may decrease or disappear. Changes of calcite or aragonite to dolomite may relate to early dolomitization, usually in restricted environments with a high pH. The necessary conditions for dolomitization are usually found in very shallow, low-energy environments, such as lagoons, where the sediments are commonly very fine.

Cementation by Overgrowths   Some kinds of particles, such as echinoderm fragments, are commonly bordered by overgrowths composed of a single crystal of calcite. If the particles are first coated with mud, have been encrusted by concentric oöid or algal coatings, or have a micritized surface, this kind of overgrowth does not normally occur. The importance of overgrowths in reducing porosity depends on the proportion of such susceptible particles in the sediment.

Cementation by Noncarbonate Minerals A number of minerals, such as evaporite minerals or chert, can be precipitated in void spaces, thus destroying original porosity. Sulfates are among the most common. They are usually precipitated when the sediments pass from an open marine setting into a more restricted environment that favors evaporation. They can also occur in subaerially exposed rocks that have undergone dissolution in the vadose zone and have then been invaded by super-saturated waters from an evaporitic environment. Chert can be deposited in limestone exposed to acid or slightly alkaline meteoric waters in which carbonates tend to dissolve.

Internal Sediment Any pores, primary or secondary, whether burrows, borings, mudcracks, or karsts, can be filled by fine-grained material carried in and deposited as internal sediment.

Pressure Solution Deterioration of reservoir quality as a result of overburden or tectonic pressures is often caused by the formation of stylolite networks accompanied by the liberation of $CaCO_3$ into solution. This $CaCO_3$ is then reprecipitated in void spaces and fissures as a calcite cement.

Cementation by Recrystallization Recrystallization during burial diagenesis increases the size of calcite crystals and changes the original crystalline texture of the limestone. Neomorphic growth by solution-precipitation produces wavy to planar crystal contacts and is associated with loss of porosity as a result of interlocking (ameboid) relationships between the neomorphic crystals. Micritization produces a fine pore network between crystals similar to that seen between fine limestone matrix material.

CONCLUSIONS

What makes or breaks porosity in carbonates is the result of many variables. In general, time is against the preservation of porosity.

Initial porosity develops in high-energy deposits in which fine-grained silt- and clay-sized particles are taken into suspension and removed into deeper water. In low-energy deposits fenestral or bird's-eye pores are created in algal-mat sediment as the organic material rots and liberates gas bubbles. The final porosity in a reservoir depends on either (1) absence of diagenesis or (2) diagenetic changes that create secondary porosity and/or destroy initial porosity.

During early subaerial alteration the most favorable zones for pore formation are those subjected to leaching by vadose waters and zones of early dolomitization. However, early diagenesis tends to reduce porosity.

During burial, favorable conditions for enhancing porosity are those linked with the development of open fractures in areas of tension, particularly in compacted and homogeneous strata.

Usually burial is linked with the destruction of the original pore network.

Karst formation is a phase of alteration where carbonate bedrock or evaporites are subjected to freshwater dissolution. Such dissolution develops megapores, including sinkholes and caves. Any carbonate rock that meteoric waters dissolve can form karstic reservoirs. However, such reservoirs are uncommon and are only preserved under exceptional conditions.

From the sedimentological and diagenetic point of view, petrographic studies lead to an understanding of the nature and importance of successive events that have affected the initial sediment. Sedimentary petrology is an important tool in reservoir studies.

REFERENCES

Archie, G. E., 1952, Classification of carbonate reservoir rocks and petrophysical considerations: *Bull Am. Assoc. Pet. Geol.,* v. 36, p. 278–298.

Asquith, G. B., 1979, *Subsurface carbonate depositional models: a concise review:* Tulsa, Okla., The Petroleum Publishing Co., 121 p.

Baille, A. D., and Vecsey, G. E., 1978, Figure 7, p. 98, and Figure 8, p. 100, in R. W. Fairbridge and J. Bourgeois, eds., *The encyclopedia of sedimentology:* Stroudsburg, Pa., Dowden, Hutchinson and Ross, 901 p.

Ball, G. R., 1968, Photographic illustration of the Arabian reservoir rock classification: Dahran, Saudi Arabia, American Institute of Mining and Metallurgical Engineers, Society of Petroleum Engineers, Reg. Tech. Symp., 2nd, Proc., p. 67–99.

Bathurst, R. G. C., 1966, Boring algae, micrite envelopes, and lithification of molluscan biosparities: *Geol. J.,* v. 5, pt. 1, p. 15–32.

Bathurst, R. G. C., 1975, *Carbonate sediments and their diagenesis,* 2nd ed.: Amsterdam, Elsevier Publishing Co., 658 p.

Bramkamp, R. A., and Powers, R. W., 1958, Classification of Arabian carbonate rocks: *Bull. Geol. Soc. Am.,* v. 69, p. 1305–1318.

Chilingar, G. V., Manon, R. W., and Rieke, H., Eds., 1972, *Oil and gas production from carbonate rocks:* Amsterdam, Elsevier Publishing Co., 408 p.

Choquette, P. W., and Pray, L. C., 1970, Geological nomenclature and classification of porosity in sedimentary carbonates: *Bull. Am. Assoc. Pet. Geol.,* v. 54, p. 207–250.

Delfiner, P. Etienne, J., and Fonck, J. M., 1972, Application de l'analyseur de textures à l'etude morphologique des réseaux poreux en lames minces: *Inst. Franc. Pétrole,* Rev., v. 22, p. 535–558.

Dunham, R. J., 1969, Vadose pisolites in the Capitan reef (Permian), New Mexico and Texas, p. 182–191, in G. M. Friedman, Ed., *Depositional environments in carbonate rocks; a symposium:* Tulsa, Okla., Society of Econ. Paleontologists and Mineralogists, Spec. Pub. No. 14, 209 p.

Dunham, R. J., 1971, Meniscus cement, p. 297–300, in O. P. Bricker, Ed., *Carbonate cements:* Baltimore and London, The Johns Hopkins University Press, 376 p.

Etienne, J., and LeFournier, J., 1967, Application des résines synthétiques colorée pour l'étude des propriétés réservoirs des roches en lames minces: *Inst. Franc. Petrole,* Rev., v. 22, p. 595–629.

Friedman, G. M., 1964, Early diagenesis and lithification in carbonate sediments: *J. Sediment. Pet.,* v. 34, p. 777–813.

Friedman, G. M., 1975, The making and unmaking of limestones or the downs and ups of porosity: *J. Sediment. Pet.,* v. 45, p. 379–398.

Friedman, G. M., 1980, Review of depositional environments in evaporite deposits and the role of evaporites in hydrocarbon accumulation: *Bull. Cent. Rech. Explor.-Prod. Elf-Aquitaine,* v. 4, p. 589–608.

Friedman, G. M., Amiel, A. J., and Schneidermann, N., 1974, Submarine cementation in reefs: examples from the Red Sea: *J. Sediment. Pet.,* v. 44, p. 816–825.

Friedman, G. M., and Sanders, J. E., 1967, Origin and occurrence of dolostones, p. 267–348, in G. V. Chilingar, H. J. Bissell, and R. W. Fairbridge, Eds., *Carbonate rocks, origin, occurence and classification:* Amsterdam, Elsevier Publishing Co., 471 p.

Friedman, G. M., and Sanders, J. E., 1978, *Principles of sedimentology:* New York, John Wiley & Sons, 792 p.

Harbaugh, J. W., 1967, Carbonate oil reservoir rocks, p. 349–399, in G. V. Chilingar, H. J. Bissell, and R. W. Fairbridge, Eds., *Carbonate rocks, origin, occurrence and classification:* Amsterdam, Elsevier Publishing Co., 471 p.

Klement, K., 1971, Genetic classification of porosity formation and destruction in carbonate rocks: *Bull. Am. Assoc. Pet. Geol.,* v. 55, p. 154.

LeRoy, L. W., LeRoy, D. O., and Raese, J. W., Eds., 1977, *Subsurface geology,* 4th ed.: Golden, Colorado, Colorado School of Mines, 941 p.

Levorsen, A. I., 1967, *Geology of petroleum,* 2nd ed.: San Francisco and London, W. H. Freeman and Co., 724 p.

Pickett, G. R., 1977, Recognition of environments and carbonate rock type identification, in *Formation evaluation manual unit II, section exploration wells:* Oil and Gas Consultants International Inc., p. 4–25.

Powers, R. W., 1962, Arabian Upper Jurassic carbonate reservoir rocks, p. 122–192, in W. E. Ham, Ed., *Classification of carbonate rocks. A symposium:* Tulsa Okla., American Association of Petroleum Geologists, Mem., 1, 279 p.

Purcell, W. R., 1949, Capillary pressures—their measurement using mercury and the calculation of permeability therefrom: *Pet. Trans.,* AIME, February p. 39–48.

Sander, N. J., 1967, Classification of carbonate rocks of marine origin: *Bull. Am. Assoc. Pet. Geol.* v. 51, p. 325–336.

Teodorovich, G. I., 1943, Structure of the pore space of carbonate reservoir rocks and their permeability as illustrated by Paleozoic reservoirs of Bashkiriya: *Dokl. Akad. Nauk S.S.S.R.,* v. 39, p. 231–234.

Teodorovich, G. I., 1958, *Study of Sedimentary Rocks:* Leningrad, Gostoptekhyzdat, 572 p.

Thomas, G. E., 1962, Grouping of carbonate rocks into textural and porosity units for mapping purposes, p. 193–224, in W. E., Ham, Ed., *Classification of carbonate rocks. A symposium:* Tulsa, Okla., American Association of Petroleum Geologists, Mem., 1, 279 p.

Waldschmidt, W. A., Fitzgerald, P. E., and Lunsford, C. L., 1956, Classification of porosity and fractures in reservoir rocks: *Bull. Am. Assoc. Pet. Geol.,* v. 40, p. 953–974.

Zenger, D. H., Dunham, J. B., and Ethington, R. L., Eds., 1980, *Concepts and models of dolomitization:* Tulsa, Okla., Society of Econ. Paleontologists and Mineralogists, Spec. Pub. No. 28, 320 p.

ADDITIONAL READING

Aschenbrenner, B. C., and Achauer, C. W., 1960, Minimum conditions for migration of oil in water-wet carbonate rocks: *Bull. Am. Assoc. Pet. Geol.,* v. 44, p. 235–243.

Aschenbrenner, B. C., and Chilingar, G. V., 1960, Teodorovich's method for determining permeability from pore-space characteristics of carbonate rocks: *Bull. Am. Assoc. Petroleum Geologists,* v. 44, p. 1421–1424.

Bricker, O. P., Ed., 1971, *Carbonate cements:* Baltimore and London, The Johns Hopkins University Press, 376 p.

Choquette, P. W., and Trusell, F. C., 1978, A procedure for making the titan-yellow stain for Mg-calcite permanent: *J. Sediment. Pet.,* v. 48, p. 639–641.

Dickson, J. A. D., 1966, Carbonate identification and genesis as revealed by staining: *J. Sediment. Pet.,* v. 36, p. 491–505.

Dunham, R. J., 1962, Classification of carbonate rocks according to depositional texture, p. 108–121, in W. E. Ham, Ed., *Classification of carbonate rocks:* Tulsa Okla., American Association of Petroleum Geologists, Mem., 1, 279 p.

Folk, R. L., 1959, Practical petrographic classification of limestones: *Bull. Am. Assoc. Pet. Geol.,* v. 43, p. 1–38.

Folk, R. L., 1974, The natural history of crystalline calcium carbonate: effects of magnesium content and salinity: *J. Sediment. Pet.* v. 44, p. 40–53.

Folk, R. L., and Land, L. S., 1974, Mg/Ca ratio and salinity: two controls over crystallization of dolomite: *Bull. Am. Assoc. Pet. Geol.,* v. 59, p. 60–68.

Friedman, G. M., 1959, Identification of carbonate minerals by staining methods: *J. Sediment. Pet.,* v. 29, p. 87–97.

Friedman, G. M., 1965, Terminology of crystallization textures and fabrics in sedimentary rocks: *J. Sediment. Pet.,* v. 35, p. 643–655.

Friedman, G. M., 1968, Geology and geochemistry of reefs, carbonate sediments and waters, Gulf of Aqaba (Elat), Red Sea: *J. Sediment. Pet.,* v. 38, p. 895–919.

Gavish, E., and Friedman, G. M., 1969, Progressive diagenesis in Quaternary to Late Tertiary carbonate sediments: sequence and time scale: *J. Sediment. Pet.,* v. 39, p. 980–1006.

Ginsburg, R. N., Marszalek, D. S., and Schneidermann, N., 1971, Ultrastructure of carbonate cements in a Holocene algal reef of Bermuda: *J. Sediment. Pet.,* v. 41, p. 472–482.

Gvirtzman, G., and Friedman, G. M., 1977, Sequence of progressive diagenesis in coral reefs, p. 357–380, in S. H. Frost, M. P. Weiss, and J. B. Saunders, Eds., *Reefs and related carbonates—ecology and sedimentology:* American Association of Petroleum Geologists, Studies in Geology, No. 4, 421 p.

Harrison, R. S., 1977, Caliche profiles: indicators of near-surface subaerial diagenesis, Barbados, West Indies: *Bull. Can. Pet. Geol.,* v. 25, p. 123–173.

James, N. P., 1972, Holocene and Pleistocene calcareous crust (caliche) profiles; criteria for subaerial exposure: *J. Sediment. Pet.,* v. 42, p. 817–836.

James, N. P., Ginsburg, R. N., Marszalek, D. S., and Choquette, P. W., 1976, Facies and fabric specificity of early subsea cements in shallow Belize (British Honduras) reefs: *J. Sediment. Pet.,* v. 46, p. 523–544.

Land, L. S., 1970, Phreatic versus vadose meteoric diagenesis in limestones: evidence from a fossil water table: *Sedimentology,* v. 14, p. 175–185.

Logan, B. W., 1974, Inventory of diagenesis in Holocene-Recent carbonate sediments, Shark Bay, Western Australia, p. 195–249, in B. W. Logan, J. F. Read, G. M. Hagan, P. Hoffman, R. G. Brown, P. J. Woods, and C. D. Gebelein, Eds., *Evolution and diagenesis of Quaternary carbonate sequences, Shark Bay, Western Australia:* Tulsa, Okla., American Association of Petroleum Geologists, Mem., 22, 358 p.

Lucia, F. J., and Murray, R. C., 1967, *Origin and distribution of porosity in crinoidal rock:* Mexico City, 7th World Petroleum Congr., Proc., v. 2, p. 409–423.

MacIntyre, I. G., 1977, Distribution of submarine cements in a modern Caribbean fringing reef, Galeta Point, Panama: *J. Sediment. Pet.,* v. 47, p. 503–516.

Matthews, R. K., 1974, A process approach to reefs and reef associated limestones, p. 234–256, in L. F. LaPorte, Ed., *Reefs in time and space:* Tulsa, Okla., Society of Economic Paleontologists and Mineralogists, Spec. Pub. No. 18, 256 p.

Milliman, J. D., 1974, *Marine carbonates:* Berlin, Springer-Verlag, 375 p.

Murray, R. C., 1960, Origin of porosity in carbonate rocks: *J. Sediment. Pet.,* v. 30, p. 59–84.

Purser, B. H., 1973, *The Persian Gulf, Holocene carbonate sedimentation and diagenesis in a shallow epicontinental sea:* Berlin, Springer-Verlag, 471 p.

Scholle, P. A., 1978, *Carbonate rock constituents, textures, cements and porosities:* Tulsa, Okla., American Association of Petroleum Geologists, Mem., 27, 241 p.

Schroeder, J. H., 1973, Submarine and vadose cements in Pleistocene Bermuda reef rock: *Sediment. Geol.,* v. 10, p. 179–205.

Shinn, E. A., 1969, Submarine lithification of Holocene carbonates in the Persian Gulf: *Sedimentology,* v. 12, p. 109–144.

Steinen, R. P., 1974, Phreatic and vadose diagenetic modification of Pleistocene limestone: petrographic observations from subsurface of Barbados, West Indies: *Bull. Am. Assoc. Pet. Geol.* v. 58, p. 1008–1024.

Thorstenson, D. C., Mackenzie, F. T., and Ristvet, B. L., 1972, Experimental vadose and phreatic cementation of skeletal carbonate sand: *J. Sediment. Pet.,* v. 42, p. 162–167.

Wardlaw, N. C., 1976, Pore geometry of carbonate rocks as revealed by pore casts and capillary pressure: *Bull. Am. Assoc. Pet. Geol.,* v. 60, p. 245–267.

Wardlaw, N. C., and Taylor, R. P., 1976, Mercury capillary pressure curves and the interpretation of pore structure and capillary behaviour in reservoir rocks: *Bull. Can. Pet. Geol.,* v. 24, p. 225–262.

Wilson, J. L., 1975, *Carbonate facies in geologic history:* Berlin Springer-Verlag, 471 p.

Appendix: Classification of Pores in Carbonate Rocks

A number of classifications are in use for carbonates which relate to their petrophysical properties. These classifications are concerned with pore geometry and its relation to texture (Waldschmidt et al., 1956), with the various kinds of pores and their interconnective networks (Teodorovich, 1958) and with the relationship between rock type and pore development (Bramkamp and Powers, 1958). Initially based on description only (Archie, 1952), classifications began to develop that took into account genetic or diagenetic origin of pores (Choquette and Pray, 1970).

CLASSIFICATIONS RELATED TO THE GEOMETRY OF PORES

The Classification of Waldschmidt et al. (1956)

The classification of Waldschmidt et al. relates pore size and infilling. Porosity of the matrix is divided into five categories: VG—very good; G—good; F—fair; P—poor; and VP—very poor.

The pores or vugs (macropores) are classified by size: VL—very large, greater than 10 mm; L—large, 4 to 10 mm; M—medium, 1 to 4 mm; and S—small, less than 1 mm; they are described according to their calcite fillings. Fractures are classified separately (Table 2-4) according to their filling and orientation. Waldschmidt's classification is simple and easy to use, but does not take reservoir heterogeneity or pore interconnections into account.

The Classifications of Levorsen (1956), Harbaugh (1967), and Klement (1971) (Table 2-5)

These classifications relate pore type and genesis. According to these classifications, which complement each other, different kinds of pores can be defined as a function of their origin and the processes that modify them, such as compaction, dissolution, recrystallization, or infilling.

Primary porosity is defined as porosity formed by depositional processes. It includes interparticle porosity as well as primary intercrystalline porosity, fenestral pores, desiccation cracks (Levorsen, 1956), framework porosity (Harbaugh, 1967), and intraparticle porosity (Klement, 1971).

Secondary porosity includes intercrystalline porosity, such as that which occurs in sucrosic dolomites in which dolomite crystals are euhedral to subhedral and slightly welded at point contacts, and porosity linked to selective dolomitization, such as that which occurs in burrows or near fractures.

Klement (1971) has classified the origin of porosity associated with dolomitization as follows:

1 Leaching resulting from the dolomitization process.
2 Volume reduction caused by the slight density difference between calcite and dolomite.
3 Preservation of primary porosity by fast diagenetic hardening.
4 Interstitial porosity created by dolomitization and subsequent recrystallization.
5 Porosity linked with carbonate dissolution, including leaching of particles or micrite.
6 Porosity linked with recrystallization (this is subdivided into leaching accompanying the recrystallization process, rearrangement of crystal fabric and preservation of primary porosity by early lithification).
7 Porosity associated with fractures.

Table 2-4 Classification of Fractures According to Waldschmidt (1956)

CLASSIFICATION OF FRACTURES IN CORES			
TYPE	**ORIENTATION** Note: B, C, & D — Single or parallel fractures		**Mineral Grains In- filling Larger or Smaller Than in Matrix**
1 OPEN	A Random	a Non-intersecting	
		b Intersecting	
	B Vertical		
	C Horizontal		
	D Inclined		
	E Intersecting	VV Vert. & Vert.	
		VH Vert. & Horiz.	
		VI Vert. & Inc.	
		HI Horiz. & Inc.	
		II Inc. & Inc.	
2 PARTLY FILLED	A Random	a Non-intersecting	x Larger y Smaller
		b Intersecting	x Larger y Smaller
	B Vertical		x Larger y Smaller
	C Horizontal		x Larger y Smaller
	D Inclined		x Larger y Smaller
	E Intersecting	VV Vert. & Vert.	x Larger y Smaller
		VH Vert. & Horiz.	x Larger y Smaller
		VI Vert. & Inc.	x Larger y Smaller
		HI Horiz. & Inc.	x Larger y Smaller
		II Inc. & Inc.	x Larger y Smaller
3 FILLED	A Random	a Non-intersecting	x Larger y Smaller
		b Intersecting	x Larger y Smaller
	B Vertical		x Larger y Smaller
	C Horizontal		x Larger y Smaller
	D Inclined		x Larger y Smaller
	E Intersecting	VV Vert. & Vert.	x Larger y Smaller
		VH Vert. & Horiz.	x Larger y Smaller
		VI Vert. & Inc.	x Larger y Smaller
		HI Horiz. & Inc.	x Larger y Smaller
		II Inc. & Inc.	x Larger y Smaller
4 CLOSED	A Random	a Non-intersecting b Intersecting	
	B Vertical		
	C Horizontal		
	D Inclined		
	E Intersecting	VV Vert. & Vert.	
		VH Vert. & Horiz.	
		VI Vert. & Inc.	
		HI Horiz. & Inc.	
		II Inc. & Inc.	

Table 2-5   The Classifications of Levorsen (1956), Harbaugh (1967), and Klement (1971)

Pore Types \ Authors	HARBAUGH	LEVORSEN	KLEMENT
PRIMARY POROSITY	INTERGRANULAR FRAMEWORK POROSITY CARBONATE MUD	INTERGRANULAR — BIOCLASTS / OÖIDS INTERCRYSTALLINE FENESTRAL PORES DESICCATION CRACKS	INTERGRANULAR INTRAGRANULAR

SECONDARY POROSITY	DISSOLUTION
	  INTRACORALLINE
	  INTERALGAL
	  SHELLS
	
	DOLOMITIZATION
	(INTERCRYSTALLINE
	POROSITY)
	
	FRACTURES
		DISSOLUTION
		
		RECRYSTALLIZATION
		
		DOLOMITIZATION
		(INTERCRYSTALLINE
		POROSITY)
		
		FRACTURES
		FISSURES
		JOINTS
			SUBAERIAL DISSOLUTION
			MOLDIC VUGS
			
			RECRYSTALLIZATION
			  —WITH DISSOLUTION
			  —INTERSTITIAL PORES
			  —PRESERVATION OF
			    PRIMARY POROSITY
			    (EARLY LITHIFICATION)
			DOLOMITIZATION
			  —WITH DISSOLUTION
			  —VOLUME DECREASE
			  —PRESERVATION OF
			    PRIMARY POROSITY
			    (EARLY LITHIFICATION)
			  —INTERSTITIAL POROSITY
			FRACTURES

CEMENTATION	CEMENTATION OF
	THE PRIMARY PORES
	AND
	REPLACEMENT
		CEMENTATION
		COMPACTION
			FIBROUS CALCITE
			
			CALCITE CRYSTALS
			
			DOLOMITE CRYSTALS
			
			INFILLING
			  —GYPSUM
			  —ANHYDRITE
			  —LIME MUD
			  —ISOLATED DOLOMITE
			CRYSTALS —OTHER

Although incomplete, these classifications are valuable because they deal with pore genesis.

Classification of Choquette and Pray (1970)

This classification relates pore types to their origin (Fig. 2-22). It also subdivides pore geometry and distribution according to their textural relationships. These relationships can be primary, say interparticle, or secondary, say moldic. Where primary sedimentary features determine the position of the pore network such a network is termed "fabric selective." Where such a relationship cannot be determined the porosity is termed "nonfabric selective." On the basis of these subdivisions, 15 pore types are distinguished, and these are fabric selective or nonfabric selective, primary (depositional) or secondary (diagenetic).

Fabric Selective

Interparticle; normally primary but can be secondary, for example dissolution of a micritic matrix.

Intraparticle; either primary, as within chambers of rudistids, or secondary because of dissolution.

Intercrystalline; usually secondary but can be primary, for example microspaces between the crystals that constitute the particles.

Moldic; secondary porosity as the result of the dissolution of particles or crystals.

Fenestral; commonly primary as the result of algal decay in algal-mat sediments and the liberation of gas bubbles. May be secondary when associated with the dissolution of unstable minerals.

Shelter; primary voids preserved beneath large particles that protect the underlying space from cementation and infilling.

Growth/framework; gaps in a boundstone framework.

Nonfabric Selective

Fracture; secondary porosity due to such various processes as compaction, tectonic forces, karstification, and solution collapse.

Channel; a secondary solution network.

Vug; secondary porosity formed by dissolution but independent of particles.

Cavern; differs from vugs only in dimension.

Fabric Selective or Not

Breccia; this kind of porosity may be the ultimate result of a channel network. It can also be primary, for example, depositional breccia, intraclast breccia.

Boring; this is produced by organisms penetrating a solid substrate.

Burrow; this kind of porosity is produced by organisms penetrating a soft substrate.

Shrinkage; most commonly due to synsedimentary desiccation or shrinkage as a result of hardening.

In addition, the following are taken into account:

1   The processes that modify pores and the effects of these processes.
2   The time of pore formation.
3   The importance of pore volume.
4   The importance of solution and infilling.

Choquette and Pray's classification is complete and detailed. It takes into account the distribution of the pore system, the origin of the pores, and the importance of alterations. To apply it, a thorough knowledge of sedimentological processes is necessary.

The Classification of Baillie and Vecsey (1978)

This classification relates porosity and the effects of diagenesis (Tables 2-6 and 2-7). The following categories can be distinguished:

1   **Primary porosity**
    (a)   Interparticle
    (b)   Interbioclastic
    (c)   Intrabioclastic
    (d)   Framework porosity
    (e)   Matrix porosity.
2   **Secondary porosity**
    (a)   Intercrystalline
    (b)   Solution voids and molds
    (c)   Fractures
3   **Destruction of porosity**
    (a)   Compaction/pressure solution
    (b)   Organic and inorganic cementation
    (c)   Secondary infilling
    (d)   Growth of dolomite

This classification is particularly useful for dolomite, but does not take into account the heterogeneity of the reservoir, nor does it explain the causes of the diagenetic alterations. It also does not deal with pore interconnections.

CLASSIFICATIONS BASED ON PORE INTERCONNECTIONS

Teodorovich (1943, 1958) (Tables 2-8 and 2-9)

This classification defines 16 pore systems based on the intercommunication between the pores, using samples of known permeability and certain observations. Permeability depends on:

1   The pore type.
2   The "usable" porosity.

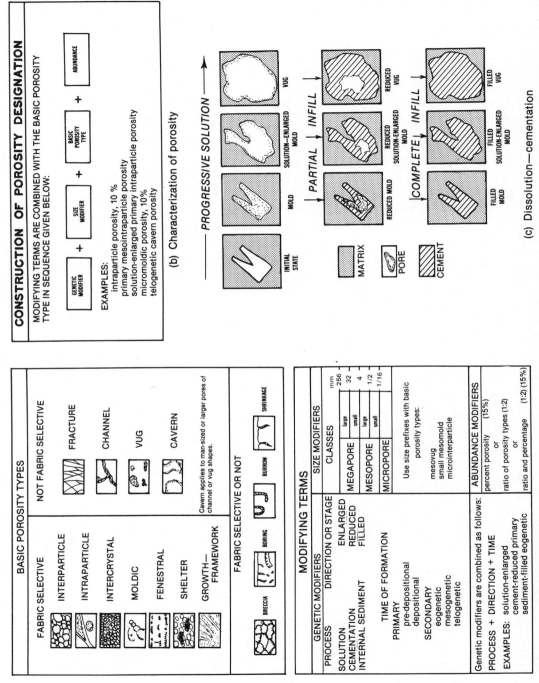

(a) Classification

(b) Characterization of porosity

(c) Dissolution—cementation

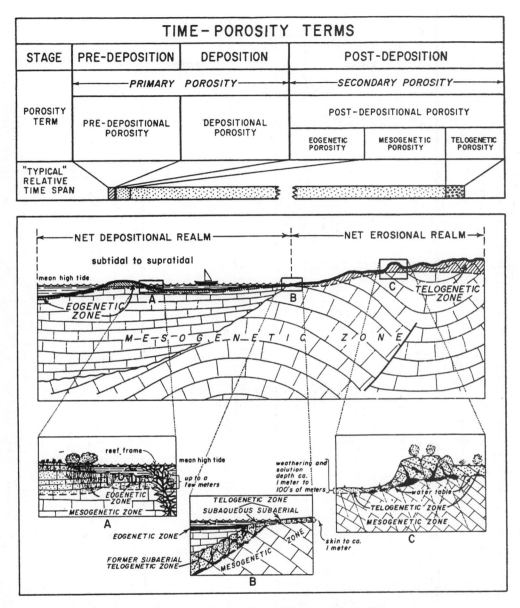

TIME–POROSITY TERMS					
STAGE	PRE-DEPOSITION	DEPOSITION	POST-DEPOSITION		
POROSITY TERM	*PRIMARY POROSITY*		*SECONDARY POROSITY*		
	PRE-DEPOSITIONAL POROSITY	DEPOSITIONAL POROSITY	POST-DEPOSITIONAL POROSITY		
			EOGENETIC POROSITY	MESOGENETIC POROSITY	TELOGENETIC POROSITY
"TYPICAL" RELATIVE TIME SPAN					

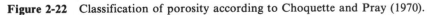

(d) Zones of diagenesis. A, B and C are the major post-depositional zones. The eogenetic and telogenetic zones extend from the surface to below the limit of subaerial weathering and erosion, the practical limit of the telogenetic zone is the water table.

Figure 2-22   Classification of porosity according to Choquette and Pray (1970).

89

Table 2-6 Classification of Porosity Development in Grain-Supported Sediments after Baillie and Vecsey, Gulf Oil Canada (Published in Fairbridge and Bourgeois, 1978, p. 98)

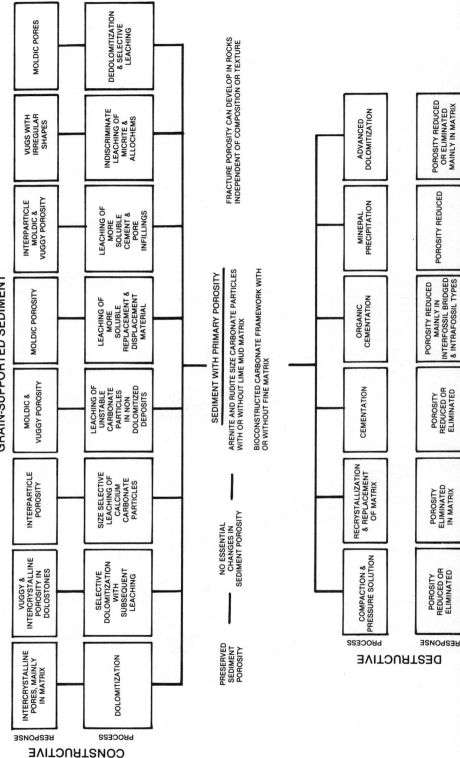

POROSITY DEVELOPMENT THROUGH DIAGENESIS

GRAIN-SUPPORTED SEDIMENT

CONSTRUCTIVE

RESPONSE / PROCESS

INTERCRYSTALLINE PORES, MAINLY IN MATRIX	VUGGY & INTERCRYSTALLINE POROSITY IN DOLOSTONES	INTERPARTICLE POROSITY	MOLDIC & VUGGY POROSITY	MOLDIC POROSITY	INTERPARTICLE MOLDIC & VUGGY POROSITY	VUGS WITH IRREGULAR SHAPES	MOLDIC PORES
DOLOMITIZATION	SELECTIVE DOLOMITIZATION WITH SUBSEQUENT LEACHING	SIZE SELECTIVE LEACHING OF CALCIUM CARBONATE PARTICLES	LEACHING OF UNSTABLE CARBONATE PARTICLES IN NON DOLOMITIZED DEPOSITS	LEACHING OF MORE SOLUBLE REPLACEMENT & DISPLACEMENT MATERIAL	LEACHING OF MORE SOLUBLE CEMENT & PORE INFILLINGS	INDISCRIMINATE LEACHING OF MICRITE & ALLOCHEMS	DEDOLOMITIZATION & SELECTIVE LEACHING

NO ESSENTIAL CHANGES IN SEDIMENT POROSITY

PRESERVED SEDIMENT POROSITY

SEDIMENT WITH PRIMARY POROSITY

ARENITE AND RUDITE SIZE CARBONATE PARTICLES WITH OR WITHOUT LIME MUD MATRIX

BIOCONSTRUCTED CARBONATE FRAMEWORK WITH OR WITHOUT FINE MATRIX

FRACTURE POROSITY CAN DEVELOP IN ROCKS INDEPENDENT OF COMPOSITION OR TEXTURE

DESTRUCTIVE

PROCESS / RESPONSE

COMPACTION & PRESSURE SOLUTION	RECRYSTALLIZATION & REPLACEMENT OF MATRIX	CEMENTATION	ORGANIC CEMENTATION	MINERAL PRECIPITATION	ADVANCED DOLOMITIZATION
POROSITY REDUCED OR ELIMINATED	POROSITY ELIMINATED IN MATRIX	POROSITY REDUCED OR ELIMINATED	POROSITY REDUCED MAINLY IN INTERFOSSIL BRIDGED & INTRAFOSSIL TYPES	POROSITY REDUCED	POROSITY REDUCED OR ELIMINATED MAINLY IN MATRIX

Table 2-7 Porosity Development in Mud or Micrite-Supported Sediments after Baillie and Vecsey, Gulf Oil Canada (Published in Fairbridge and Bourgeois, 1978, p. 100)

POROSITY DEVELOPMENT THROUGH DIAGENESIS

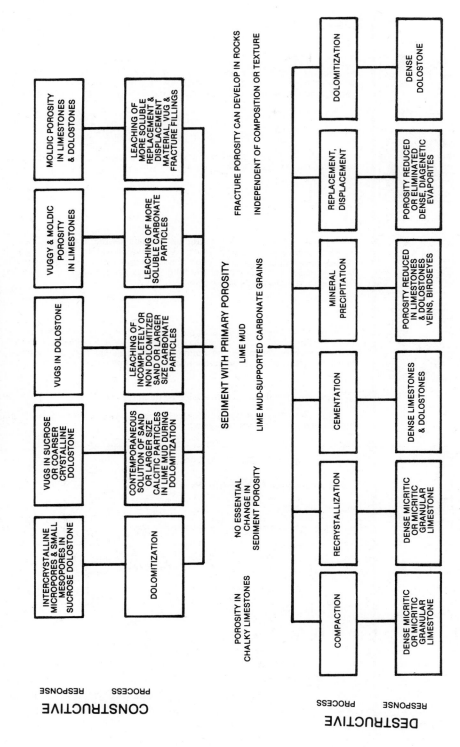

Table 2-8   Classification of Carbonate Porosity after Teodorovich (1943)

Type I — The pore spaces of this type consist of pores and of rather isolated more or less narrow conveying canals. Commonly the narrow canals (inner diameter 0.01-0.005 mm) which connect the pores of this first type are not visible in a thin section. If the minimum diameter of a canal is larger, however, the canal can be detected in a thin section.

Type II — The communicating ducts of the pore spaces of the second type consist merely of constrictions in the pore spaces which become wider and pass gradually into the pores proper.

Type III — The third type of structure is characterized by the presence of pores connected by finely porous broad canals which are observed in a thin section in the form of branches. Some conveying canals may consist of coarser pores which sharply increases the permeability The pore-space configuration of the third type is usually found in dolomites; less commonly, it is observed in dolomitic limestones.

Type IV — The fourth type of structure of pore spaces is characterized by a system of pores distributed between the grains and near the grains of the main mass of dolomite rock or of its cement, reflecting the outlines of the greatest part of these grains (intergranular pores). The interrhombohedral porosity in dolomites serves as an excellent example.

Type V — The pore space is formed by fractures.

Type VI — The pore space is characterized by two or more types of pore space configuration (mixtures of types I through V).

Legend:

(hatched)	Pores clearly observable in thin sections.
(canal)	Fine conveying canals between pores.
(stippled)	Intergranular pores.
(branches)	Fine and extremely fine-grained conveying branches.

3   The average size of the pores.
4   The shape of the pores.

Each pore type is subdivided into classes and the product of the numerical coefficient given to the pore class defines the permeability. This is the most detailed classification of its kind. Its accuracy has been confirmed by Aschenbrenner and Chilingar (1960) as about 13% for pore evaluation. It is a descriptive classification, but does not permit extrapolations.

Table 2-9   Relationship between Kinds of Pores as Observed in Thin Sections and Measurements of Porosity (after Teodorovich, 1960)

(a) Empirical coefficient A for pore shape.

Pore Space Type	Characteristic of Subtype (As Seen in Thin Section.)	Empirical Coefficient A
I	With very narrow conveying canals (av. diameter $\approx$ 0.01 mm), usually not visible in thin section under the petrographic microscope using normal range of magnification	2
	With rare relatively wide canals (avg. diameter $\geq$ 0.02 mm) visible in thin sections	4
	With few relatively wide canals, visible in thin section	8
	With many relatively wide canals, visible in thin section, or with few wide canals (avg. diameter $\geq$ 0.04 mm.)	16
	With abundant wide canals or few to many very wide conveying canals	32
II	With poor porosity, the pores being relatively homogeneous in size and distribution	8
	With good porosity and (or) porosity ranging from poor to good:	
	pores being of different size	16-32
	pores being vuggy and irregular in outline	32-64
III	With very poor porosity inside the conveying canals	6
	With poor porosity inside the conveying canals	12
	Conveying canals finely porous	24
IV	With interconnected pore space between the rhombohedral grains	10
	With interconnected pore space between subangular-subrounded grains	20
	With interconnected pore space between rounded to well rounded grains	30

(b) Empirical coefficient B for porosity.

Effective Porosity		
Descriptive Term	Limits in Per Cent	Empirical Coefficient B
Very porous	> 25	25-30
Porous	15-25	17
Moderately porous	10-15	10
Pores abundant	5-10	2-5
Pores present	2- 5	0.5-1.0
Some pores present	< .2	0

(c) Empirical coefficient C for pore size.

Descriptive Term	Maximum Size of Pore (mm.)	Empirical Coefficient C
Large vugs	> 2.00	16
Medium to large vugs	0.50-2.00	4
Medium pores	0.25-1.00	2
Fine to medium pores	0.10-0.50	1
Very fine to fine pores	0.05-0.25	0.5
Very fine pores	0.01-0.10	0.25
Pinpoint to very fine pores	< 0.10 and in part < 0.01	0.125
Mostly pinpoint porosity	0.03 and in part < 0.01	0.0625

(d) Empirical coefficient D for pore shape.

Descriptive Term	Empirical Coefficient D
More or less isometric pores	1
Elongate pores	2
Very elongate pores or pores arranged in bands with emanating conveying canals	4

(e) Comparison between calculated and measured permeability for different pore space.

Pore Space Type	Measured Permeability md.	Calculated Permeability md.	Difference md.
I	62.3	91.0	28.7
I	8.6	10.0	1.4
I	8.7	10.5	1.8
I	7.4	8.0	0.6
II	61.9	90.0	28.1
II	51.3	62.5	11.2
I	13.1	12.7	0.4
II	88.7	120.0	31.3
I	14.5	15.0	0.5
II	50.5	60.0	9.5
II	33.7	40.0	6.3
I	104.6	120.0	15.4
II	50.54	61.1	10.56
II	163.9	150.0	13.9
I	101.9	110.0	8.1
I	52.87	34.0	18.87
I	15.8	15.0	0.8
II	115.0	156.0	41.0
I	795.7	800.0	4.3

93

CLASSIFICATIONS BASED ON THE RELATIONSHIPS BETWEEN ROCKS AND POROSITY

Archie (1952) (Table 2-10)

Archie based his classification on the relationships between pore size and the physical texture of the rock. The percentage of visible porosity is also determined.

Archie's classification is simple and practical, especially for fieldwork, and is easily handled by nonspecialized personnel. It is useful only for preliminary examination, because it does not contain sufficient detail for dealing with heterogeneous reservoirs.

Powers (1962) (Table 2-11)

Powers provides a classification in which rocks are grouped according to their state of alteration and their particle or crystal size. Powers distinguishes two classes—sediments in which the fabric has been visibly altered and sediments that show no

Table 2-10   Classification of Archie (1952)

Texture of Matrix	Macroscopic Appearance	Microscopic Appearance 10X to 15X
Type I Compact Crystalline	— hard, dense — sharp edges and smooth faces on breaking	— matrix composed of tightly interlocking crystals, no visible pore space — edges of breaks not clean
Type II Chalky	— dull, earthy, siliceous or argillaceous — crystals less tightly interlocking than type 1 — composed of fine-grained particles or tests of marine organisms	— crystals joining at different angles, extremely fine grained, appears chalky — grain size 0.05 mm
Type III Granular or Saccharoidal	— sandy or sugary appearance — size of crystals or granules classed as: 0.05 mm: very fine 0.1 mm: fine 0.2 mm: medium 0.4 mm: coarse	— crystals interlocking at different angles but considerable porosity between crystals — oölitic and other particle textures are of this type

Classification of Visible Pores

Class A: No visible porosity under 10X microscope or where pore size is less than 0.01 mm in diameter
Class B: Visible porosity > 0.01 and < 0.1 mm
Class C: Visible porosity > 0.1 and < size of cuttings
Class D: Visible porosity is shown by secondary crystal growth on cutting faces
          -is shown by fractures or solution channels
          -pore size > cuttings

Classification of Visible Pore Frequency

Description	Frequency percent of surface covered by pores
Possibly excellent	20%
Possibly good	15%
Possibly fair	10%
Possibly poor	5%

Table 2-11 Classification of Limestones According to Powers (1962)

ORIGINAL TEXTURE	ORIGINAL TEXTURE NOT VISIBLY ALTERED (except by cementation) — ORIGINAL PARTICLE TYPE						ORIGINAL TEXTURE ALTERED — MODERATELY		ORIGINAL TEXTURE ALTERED — STRONGLY			ORIGINAL TEXTURE OBLITERATED
	MORE THAN 25% SKELETAL[2] REMAINS	MORE THAN 25% AGGREGATE[3] GRAINS	MORE THAN 25% OOLITHS	MORE THAN 25% DETRITUS[4] FROM OLDER LS.	10-50% NONCARBONATE SAND	10-50% NONCARBONATE MUD	WEAKLY DEVELOPED CALCITE MOSAIC (<10% DOLOMITE)	MORE THAN 10% DISCRETE DOLOMITE RHOMBS	STRONGLY DEVELOPED CALCITE MOSAIC (<10% DOLOMITE)	25-75% INTERLOCKING DOLOMITE	MORE THAN 75% DOLOMITE WITH RELIC TEXTURE	MORE THAN 75% DOLOMITE
APHANITIC LIMESTONE (Lime mud with less than 10% sand- or gravel-size clastic carbonate grains.)	*Origin of mud-size particles generally indeterminate* (chalk)				*Sandy*	*Impure* (marl)	*Partially recrystallized*	*Partially dolomitized*	*Strongly recrystallized*	*Strongly dolomitized*	*Aphanitic dolomite*	**CRYSTALLINE DOLOMITE**
CALCARENITIC LIMESTONE (More than 10% sand- or gravel-size clastic carbonate grains set in more than 10% original mud-size matrix.)	*Skeletal[2]* calcarenitic limestone	*Aggregate[3]* calcarenitic limestone	*Oolite* calcarenitic limestone	*Detrital[4]* calcarenitic limestone							*Calcarenitic dolomite*	
CALCARENITE (Sand-size clastic carbonate grains dominant; contains less than 10% original mud-size matrix.)	*Skeletal* calcarenite (coquina)	*Aggregate* calcarenite	*Oolite* calcarenite	*Detrital* calcarenite							*Calcarenitic dolomite*	
COARSE CARBONATE (Gravel-size clastic carbonate grains dominant; contains less than 10% original mud-size matrix.)	Coarse *skeletal* carbonate (coquina)	Coarse *aggregate* carbonate	Coarse *oolite* carbonate (pisolite)	Coarse *detrital* carbonate							*Coarse carbonate dolomite*	
RESIDUAL ORGANIC (Rocks composed dominantly of attached reef-building organisms still in growth position.)	*Residual algae, residual coral,* etc.										*Residual organic dolomite*	

Where recrystallization (not involving dolomite) obliterates original texture, rock is termed CRYSTALLINE LIMESTONE

1. Chart shows main rock groups in UNDERLINED CAPITAL LETTERS; modifiers of main rock groups in *italics*. For example: sandy, oolite calcarenite or impure, partially dolomitized, foraminiferal calcarenitic ls.
2. Skeletal is used here as a general modifier; specific modifiers include: foraminiferal, crinoidal, algal, coral, etc.
3. Aggregate grain is a general term used for all discrete, penecontemporaneous, sand- and gravel-size grains formed on the sea floor by (1) the tearing-up, movement, and redeposition of fragments of semi-consolidated bottom sediment or (2) the aggregation of finer particles by cementation (Illing, 1954). Specific types are angular aggregate, pellet (rounded aggregate), faecal pellet, and algal nodule.
4. The detrital carbonates contain more than 25% grains which have been (a) formed by the mechanical disintegration of older, well-consolidated limestones, (b) transported and, (c) redeposited as part of a younger sediment. Detrital grains are commonly distinguished by their inappropriate fossil content, color, lithology, etc.

Table 2-13 Classification of Carbonates Based on Particle Size, Origin, and Degree of Recrystallization.

GRAIN SIZE	CLASS 1 — SEDIMENTARY FABRIC NOT VISIBLY ALTERED						CLASS 2 — SEDIMENTARY FABRIC ALTERED				CRYSTAL SIZE
	BIOCLASTS (More than 25%)	PELLETS (More than 25%)	OÖLITHS (More than 25%)	COLLOCLASTS (More than 25%)	CARBONATE DETRITUS (More than 25%)	TERRIGENOUS DETRITUS (10 to 50%)	MODERATELY ALTERED (Less than 75% xystols)	STRONGLY ALTERED (More than 75% xystols)	RELIC TEXTURE ONLY	ORIGINAL TEXTURE OBLITERATED	
CALCILUTITE [a] (<.004 mm.)	CALCILUTITE (Origin of mud-size particles is commonly indeterminate. If they are identifiable, specify, e.g. coccolithic chalk.)					MARL & IMPURE CALCILUTITE & CALCISILTITE	PARTLY RECRYSTALLIZED OR DOLOMITIZED CALCILUTITE		MICROCRYSTALLINE LIMESTONE OR DOLOMITE WITH RELIC...TEXTURE	MICROCRYSTALLINE LIMESTONE OR DOLOMITE	MICRO-CRYSTALLINE (<.004 mm)
CALCISILTITE [a] (.004 to .0625mm.)	BIOCLASTIC CALCISILTITE (some chalk)	PELLETIC CALCISILTITE	OOLITIC CALCISILTITE	COLLOCLASTIC CALCISILTITE	DETRITAL CALCISILTITE		PARTLY RECRYSTALLIZED OR DOLOMITIZED CALCISILTITE	Strongly altered carbonates should be classified according to their crystal size. The name, however, should reflect the original nature of the rock. Modifiers are used to show the type of alteration. A qualifying phrase should be added to indicate the presence of solution phenomena, e.g., STRONGLY DOLOMITIZED MEDIUM CRYSTALLINE OOLITIC CALCARENITE WITH SOME OÖLITHS LEACHED AND OTHERS REPLACED BY SPAR CALCITE CEMENT.	FINELY CRYSTALLINE LIMESTONE OR DOLOMITE WITH RELIC...TEXTURE	FINELY CRYSTALLINE LIMESTONE OR DOLOMITE	FINELY CRYSTALLINE (.004 to .06mm)
LUTACEOUS CALCARENITE [b] (.004 to .0625mm.)	BIOCLASTIC LUTACEOUS CALCARENITE	PELLETIC LUTACEOUS CALCARENITE	OOLITIC LUTACEOUS CALCARENITE	COLLOCLASTIC LUTACEOUS CALCARENITE	DETRITAL LUTACEOUS CALCARENITE	ARGILLACEOUS CALCARENITE	PARTLY RECRYSTALLIZED OR DOLOMITIZED LUTACEOUS CALCARENITE		MEDIUM CRYSTALLINE LIMESTONE OR DOLOMITE WITH RELIC...TEXTURE	MEDIUM CRYSTALLINE LIMESTONE OR DOLOMITE	MEDIUM CRYSTALLINE (.06 to .25mm)
CALCARENITE [c] (.0625 to 2mm.)	BIOCLASTIC CALCARENITE	PELLETIC CALCARENITE	OOLITIC CALCARENITE OR OOLITE	COLLOCLASTIC CALCARENITE	DETRITAL CALCARENITE	AREMACEOUS CALCARENITE	PARTLY RECRYSTALLIZED OR DOLOMITIZED CALCARENITE		COARSELY CRYSTALLINE LIMESTONE OR DOLOMITE WITH RELIC...TEXTURE	COARSELY CRYSTALLINE LIMESTONE OR DOLOMITE	COARSELY CRYSTALLINE (.25 to 1mm)
CALCIRUDITE (2 to 64mm.)	The name of the organisms making up the abundant bioclasts should be substituted when identifiable, e.g. "Algal," "Coralline," "Orbitoidal," "Nummulitic."	PELLETIC CALCIRUDITE	PISOLITE	COLLOCLASTIC CALCIRUDITE	DETRITAL CALCIRUDITE	COBBLE CALCIRUDITE	PARTLY RECRYSTALLIZED OR DOLOMITIZED CALCIRUDITE		VERY COARSELY CRYSTALLINE LIMESTONE OR DOLOMITE WITH RELIC...TEXTURE	VERY COARSELY CRYSTALLINE LIMESTONE OR DOLOMITE	VERY COARSELY CRYSTALLINE (>1 mm)
COARSE CARBONATE (>64mm.)					LIMESTONE CONGLOMERATE	CONGLOMERATIC LIMESTONE	PARTLY RECRYSTALLIZED OR DOLOMITIZED COARSE CARBONATE		RECRYSTALLIZED COARSE CARBONATE WITH RELIC...TEXTURE		
STATOBIOLITH [d]	ALGAL OR CORAL REEF OR BIOHERM, MOLLUSCAN, CRINOID, OR ALGAL BIOSTROME OR BANK. (Not easily determinable in subsurface) OR ALGAL STATOBIOLITH, ETC.						PARTLY RECRYSTALLIZED OR DOLOMITIZED STATOBIOLITH		CRYSTALLINE CARBONATE WITH INDICATIONS OF ORIGINAL STATOBIOLITHIC NATURE.		

a. Not more than 10 per cent coarser detritus in a fine-grained matrix.

b. Calcarenite with more than 10 per cent mud or silt-size matrix.

c. Calcarenite with less than 10 per cent mud or silt-size matrix.

d. Rock composed mainly of attached reef or shoal-building organisms in their position of growth.

1. BIOCLASTS — All debris made up of the supporting or protective structures of animals and plants, whole or in fragments. Distinction must not be made between bioclasts consisting of the whole tests of small benthonic or planktonic organisms, and those composed by the entire or fragmentary remains of sessile organisms, for their genetic significance may be very different.

2. PELLETS — Elongated ellipsoidal, or rod-shape grains without internal structure made up almost exclusively of calcilutite. When transported they comprise a portion of Folk's "Intraclasts."

3. OÖLITHS — Ovate or spherical grains less than 2 mm. in diameter made up of distinctly laminated radial or concentric layers. Larger grains of the same type are called pisoliths. If a nucleus is present making up half or more of the diameter, the term "Superficial Oolith" may be used. A rock made up of ooliths is an oolite.

4. COLLOCLASTS — Autochthonous or para-autochthonous penecontemporaneous weakly cemented aggregates of calcilutite or calcisiltite with complex internal structure, but lithologically like the rock in which they are found. If they have demonstrably been transported they comprise a portion of Folk's "Intraclasts."

5. CARBONATE DETRITUS — Grains and debris derived from a pre-existing lithified carbonate rock.

6. TERRIGENOUS DETRITUS — Grains and debris derived from pre-existing rock other than carbonates.

CLASSIFICATION OF CARBONATES BASED ON PARTICLE SIZE, ORIGIN, AND DEGREE OF RECRYSTALLIZATION

Table 2-14 Carbonate Textures as Related to Porosity-Permeability Ranges

CARBONATE TEXTURES AS RELATED TO POROSITY-PERMEABILITY RANGES

CLASS 1 — ORIGINAL FABRIC NOT VISIBLY ALTERED
CARBONATE ONLY OR UP TO 50% TERRIGENOUS CLASTICS

CLASS 2 — ORIGINAL SEDIMENTARY FABRIC ALTERED

PARTICLE GRADE	COMPACTED ONLY	WEAKLY CEMENTED	MODERATELY CEMENTED	STRONGLY CEMENTED	MODERATELY (Recrystallized or dolomized)	STRONGLY (Recrystallized or dolomized)
CALCILUTITE (<.004mm)	Ø - M to VH; K - M to VL	Ø - M to VL; K - VL to Nil	Ø - VL to Nil; K - VL to Nil	—	Ø - L to M; K - L to VL (dolomite 6-78%)	Ø - H (78 - 90% dol.) to VL (>95% dol.); K - H (78 - 90% dol.) to VL (>95% dol.)
CALCISILTITE (.004 to .0625mm)	Ø - M to VH; K - M to VL	Ø - M to VL; K - L to VL	Ø - VL to Nil; K - VL to Nil	—	Ø - M to L; K - M to VL (dolomite 6-78%)	Ø - H (78 - 90% dol.) to VL (>95% dol.); K - H (78 - 90% dol.) to VL (>95% dol.)
LUTACEOUS CALCARENITE [a]	Ø - M to VH; K - M to VL	Ø - M to VL; K - L to Nil	Ø - VL to Nil; K - VL to Nil	—	Ø - H to VL; K - M to VL	
CALCARENITE (.0625 to 2mm)	Ø - M to VH; K - M to VH	Ø - M to H; K - M to H	Ø - M to VL; K - M to VL	—	Ø - H to VL; K - M to VL	
CALCIRUDITE (2 to 64mm)	Ø - M to VH; K - L to VH	Ø - M to H; K - M to H	Ø - M to Nil; K - M to Nil	—	Ø - M to Nil; K - M to Nil	
COARSE CARBONATE (>64mm)	Ø and K variable but commonly M to H	Ø and K variable but commonly M to L	Ø and K variable but commonly M to VL	—	Ø - H to Nil; K - H to Nil	
STATOBIOLITH [b]	Ø and K variable but commonly M to H	Ø and K variable but commonly M to L	Ø and K variable but commonly M to VL	—	Ø and K variable but commonly M to VL	

For STRONGLY CEMENTED: Porosity and Permeability Very Low to Nil.

RELIC TEXTURE ONLY — Ghosts of pre-existing textures still present do not affect porosity or permeability. If recrystallization is accompanied by preferential solution so that voids replace ooliths or fossils, moderate porosity and permeability may be developed. Sparry dolomitization, replacing calcarenite, etc. may yield Ø and K values as indicated under calcilutite. Strongly recrystallized rocks may contain small or large vugs which increase porosity. These are considered to be solution phenomena.

ORIGINAL TEXTURE OBLITERATED — In examples where they resemble similar rocks showing relic textures, most dolomite showing no trace of the original rock texture, together with similar limestone made up of sparry calcite formed by grain growth, can more or less confidently be assigned to this class. Other carbonates with no remaining indication of origin and made up of crystals of large size are not assignable genetically, but are put in this class arbitrarily. Porosity and permeability usually are very low, but may exist because of solution or intercrystalline voids. Intergranular as well as intercrystalline porosity may exist rarely in dolomite with relic texture as indicated in the column on the left. This is found commonly with evaporites and may truly represent a distinct category of dolomite, deposited as such.

CRYSTAL GRADE
- MICROCRYSTALLINE (<.004mm)
- FINELY CRYSTALLINE (.004 to .06mm)
- MEDIUM CRYSTALLINE (.06 to .25mm)
- COARSELY CRYSTALLINE (.25+mm)
- VERY COARSELY CRYSTALLINE (>1mm)

FRACTURE & FISSURE POROSITY — Fissures formed by solution along bedding planes or fractures are not uncommon in limestone, but these and other karstic phenomena are more numerous in rocks lying well above the water table. Where present in oil wells, fissures can be very efficient collecting locales for oil from finer-grained rocks around the fissure. Many oil fields exist because of fracturing in rocks which at least in part are too fine-grained to yield oil otherwise. Every effort must be made to determine the orientation and number of fractures, as well as their dimensions, and the number filled with calcite or other mineral.

a. Calcarenite with more than 10 per cent mud or silt-size matrix.

b. Rocks composed mainly of attached reef- or shoal-building organisms in their position of growth.

KEY

Ø = Porosity
K = Permeability

Nil = 0

	Porosity (Ø)	Permeability (K)
VL - Very Low	Ø = .1 - 3%	K = .1 - 10 md.
L - Low	Ø = 3 - 8%	K = 10 - 50 md.
M - Moderate	Ø = 8 - 16%	K = 50 - 300 md.
H - High	Ø = 16 - 25%	K = 300 - 1000 md.
VH - Very High	Ø = > 25%	K = > 1000 md.

The ranges of porosity and permeability indicated in the several categories of rock are those most likely to be encountered, but may not include the entire range of these properties in the category.

Table 2-15 Classification According to Ball (1968)

	Original Texture Not Visibly Altered Except by Cementation	Original Texture Altered					Original Texture Obliterated	Visual Porosity Classes
		Moderately		Strongly				
	<10% dolomite	<10% dolomite	>10% topic porphyrotopic dolomite	>10% dol	10 < dol < 75%	>75% dol with relicts	>75% dol.	
Group 1 Fine Limestone (<10% sand-sized or coarser clastic carbonate fragments)	Fine-ground Limestone	Partially recrystallized fine-grained limestone	Partially dolomitized fine-ground limestone	Crystalline limestone	Dolomitic limestone	Calcitic dolomite		(A): Porous and fairly cemented
Group 2 Calcarenitic Limestone (>10% sand-sized or coarser clastic carbonate fragments Fine matrix 10%)	Calcarenitic Limestone	Partially recrystallized calcarenitic limestone	Partially dolomitized calcarenitic limestone	Crystalline calcarenitic limestone	Dolomitic calcarenitic limestone	Calcarenitic dolomite	Group 4: Crystalline dolomite	(B): Compacted and/or moderately cemented
Group 3 Calcarenite Sand-sized clastic Carbonate Fragments Fine matrix <10%	Calcarenite	Partially recrystallized calcarenite	Partially dolomitized calcarenite	Strongly recrystallized calcarenite	Dolomitic calcarenite	Calcarenite dolomite		(C): Very compacted and/or strongly cemented
Group 5 Coarse Limestone Coarser than sand-sized clastic carbonate fragments Fine matrix — <10%	Coarse Clastic Carbonate	Partially recrystallized coarse clastic carbonate	Partially dolomitized coarse clastic carbonate	Strongly recrystallized coarse clastic carbonate	Dolomitic coarse clastic carbonate	Coarse clastic carbonate dolomite		

Special Rock Types

Group 1	Fine-grained limestone with small visible voids
	Compact, fine-grained limestone with small, visible voids

Group 3	Strongly recrystallized calcarenite (voids in place of oöids)
Group 5	Strongly recrystallized, coarse clastic carbonate (voids in place of fragments)

• Clastic carbonate fragments include sand-sized and coarser than sand-sized particles irrespective of origin (bioclasts, oöids, lithclasts . . . .)

alteration. Particles are classified according to their size (six divisions) and the organisms that formed them. The crystal size in altered sediments is also taken into account.

Although this classification is very detailed, it was developed for use in a particular reservoir setting and is not generally applicable.

Thomas (1962)

Thomas has developed a classification based on carbonate matrix and its dolomitized equivalents (Table 2-12). The rocks are further subdivided according to their components and the presence or absence of cementing material. The pore morphology is characterized by taking into account the particle size of the matrix, recrystallization, and cementation.

Sander (1967)

Sander subdivides rocks according to textural criteria and state of recrystallization (intensity of recrystallization, crystal size) (Tables 2-13 and 2-14).

Sander's classification for dolomites is quite elaborate and can be applied to deposits with interparticle and intercrystalline porosity. This classification does not take heterogeneous reservoirs into account and appears to apply only in special cases.

Ball (1968)

Ball provides a classification that is modified after Bramkamp and Powers (1958) (Table 2-15). The classification includes a visual estimation of porosity, but does not deal specifically with reservoir heterogeneity or pore origin.

Case Histories of
Carbonate Reservoirs

3

The search for potential reservoirs in carbonate sequences must take into account the many different factors that may be involved in reservoir formation. These factors include both depositional and diagenetic processes.

An integrated approach should be adopted for the study of carbonate sedimentary sequences. This involves studying the sequence at all levels of detail, from that of its overall tectonic setting to details of individual particles in specific beds. The study should ideally include details of the large-scale geometry of the sediment bodies that make up the sequence, their relationships to adjacent sediment bodies, their large- and small-scale internal structures, fabric, texture, composition, including floral and faunal components, and diagenetic alterations. In defining the dynamics of the depositional basin in which the sequence occurs, recognition of both large- and small-scale regressive and transgressive events is important.

Facies models are useful aids in interpreting carbonate sequences. These models are developed based on a synthesis of modern carbonate environments and studies of ancient carbonate sequences. They provide a general summary of a specific depositional environment, including a general vertical and lateral sequence of sediments, their internal structures and compositions, and an indication of the kinds of diagenetic changes that may be expected. Most sequences show some variation from an idealized facies model, but the model is still of use as the basis for interpretation, especially when the data available to the geologist are limited.

The idea of sedimentary associations is thus a very important concept in the study of carbonates. Sequences composed of certain kinds of sediments occurring together indicate particular environmental settings. Recognition of these associations through the application of general facies models greatly aids in environmental interpretations.

The distribution of porosity in carbonate sequences defines the sites of potential reservoirs. This porosity may be primary, related to the environment of deposition, or secondary, related to later diagenetic alterations.

Primary porosity is principally a function of the energy in the environment of deposition. In general, carbonate sediments deposited in high-energy environments, such as reefs or shoals, have good primary porosity. In shoals this porosity is the result of the removal of lime mud from the high-energy depositional environment by waves

Example 1 101

and currents, leaving coarser fragments with high interparticle and intraparticle porosities.

Secondary porosity is the result of diagenetic alterations, most commonly exposure to freshwater or fracturing during tectonic movements. Diagenetic processes may completely destroy all the original primary porosity in a sequence and overprint a secondary porosity which is totally unrelated to original sedimentary features.

Three kinds of carbonate reservoirs are common in the rock record.

1 Carbonate reservoirs whose presence is controlled by the primary porosity in a carbonate sequence. These reservoirs occur in high-energy deposits such as reefs, oöid shoals, and bioclastic mounds or shoals. In an idealized shelf profile, these reservoirs are associated with the development of a barrier across the shelf or in coarse gravity-displacement deposits that occur at the toe of the slope and in the basin.

2 Reservoirs controlled by secondary porosity. This porosity is commonly the result of exposure to freshwater or of fracturing in rocks during tectonic movements. Reservoirs formed during freshwater diagenesis are commonly found in shallow-water carbonate deposits. These deposits have a greater chance of being elevated above sea level during periods of uplift, eustatic fall in sea level, or relative increase in the rate of deposition. Those formed by tectonic fracturing are generally unrelated to original depositional setting.

3 Reservoirs formed by the syngenetic dolomitization of peritidal deposits. These reservoirs are formed when peritidal sediments are exposed to hypersaline waters conducive to the formation of dolomite. The dolomitization produces intercrystalline porosity in otherwise tight lime muds.

EXAMPLE 1. A PERITIDAL DOLOMITE RESERVOIR—THE ORDOVICIAN RED RIVER FORMATION, WILLISTON BASIN, MONTANA, U.S.A.

The Red River Formation in the Cabin Creek field, Montana, is an Upper Ordovician sequence of peritidal carbonates. The sequence is composed of a number of shallowing-upward depositional cycles that pass from the shallow marine into the supratidal environment. Reservoirs in the Red River Formation are primarily controlled by the occurrence of dolomite. This dolomite is best developed at the top of each shallowing-upward cycle and is believed to be the result of exposure of the sediments to hypersaline waters in the supratidal zone. (Ruzyla, 1980; Ruzyla and Friedman, 1981).

Geological Setting

The Williston Basin is a wide, circular basin with indistinct boundaries, formed on the southern side of the Canadian shield (Fig. 3-1). The basin contains sedimentary rocks that range in age from Ordovician to Tertiary. The Lower Paleozoic sequence in the Williston Basin is composed mainly of carbonates that were deposited in a shallow epeiric sea and show evidence of cyclic deposition in environments ranging from shallow marine to supratidal (Roehl, 1967). The sequence contains bioclastic and argillaceous limestones, dolomites, and evaporites. In the Cabin Creek field, oil production is from three of these carbonate units, the Upper Ordovician Red River

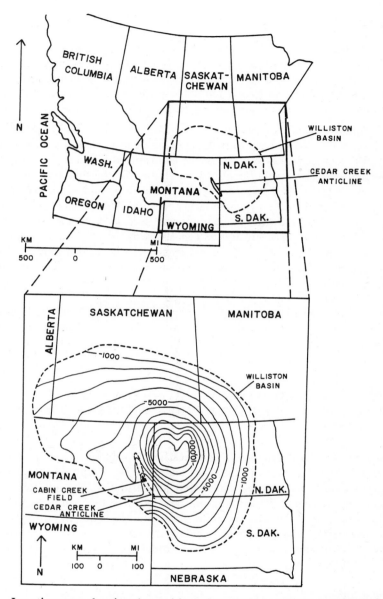

Figure 3-1 Location map showing the position of the Williston Basin and the Cedar Creek Anticline. Structure contours on the top of the Red River Formation are in feet (from Ruzyla and Friedman, 1981, p. 409).

Example 1 103

Formation, the Stony Mountain Formation, and the Silurian Interlake Formation (Fig. 3–2). The Red River Formation produces hydrocarbons from three dolomitic horizons, the U2, U4, and U6, which occur in the upper 46 m of the formation, interbedded with nonproductive limestone units, the U1, U3, and U5 (Fig. 3-3).

Facies and Environments

The Red River Formation is from 91 to 183 m thick in the Williston Basin but production is restricted to the upper 46 m. The lower part of the Red River Formation is a nonporous marine limestone that has been partially dolomitized (Fuller, 1961). The upper 46 m of the formation is a cyclical sequence composed of three limestone and three dolostone units indicative of environments ranging from subtidal to supratidal.

The subtidal lithologies in the Red River Formation consist of mottled dolomitic biomicrites that contain fragments of echinoderms, brachiopods, bryozoans, trilobites, molluscs, ostracodes, and gastropods, together with less common solitary corals and stromatoporoids. These deposits are micrite supported, poorly bedded, and heavily burrowed and bioturbated and indicate deposition in a low-energy environment.

The intertidal deposits consist of dolomitic biomicrites that show evidence of deposition under higher-energy conditions. These deposits show graded bedding, cross-bedding, and the presence of ripple marks, imbricate shells, and conglomerates. In addition to the skeletal particles found in the subtidal deposits, the intertidal sediments contain oncolites, intraclasts, peloids, and oöids.

The supratidal sediments within the Red River Formation show evidence of desiccation cracks and include intraclast breccias, laminated dolomites, solution-collapse breccias, anhydrite, and organic-rich black muds. These supratidal deposits show evidence of vadose leaching through the occurence of empty molds of anhydrite and gypsum, solution-collapse breccias, and corrosion around remnant anhydrite crystals. Some of the anhydrite leached from the supratidal deposits has been reprecipitated within the underlying intertidal and subtidal sediments.

Dolomitization

Dolomitization within the upper Red River Formation is considered to have occurred in two stages. The first stage consists of syngenetic or penecontemporaneous dolomitization of the supratidal deposits in response to the hypersaline conditions in their environment of deposition; the second stage is the gradual dolomitization of the underlying intertidal and subtidal deposits by the downward penetration of these hypersaline waters (Ruzyla, 1980; Ruzyla and Friedman, 1980).

Dolomitization of the supratidal sediments is almost complete. The rocks are laminated and peloidal, finely crystalline, hypidiotopic, and microcrystalline xenotopic dolomites. Vugs within the dolomites produced by the vadose leaching of evaporite minerals have not been filled by the dolomite rhombs, implying that the dolomitization of the sediments occurred before vadose leaching. The hypersaline conditions indicated by the presence of evaporites within the supratidal sediments is in keeping with modern occurrences of syngenetic dolomite (Zenger, Dunham, and Ethington, 1980).

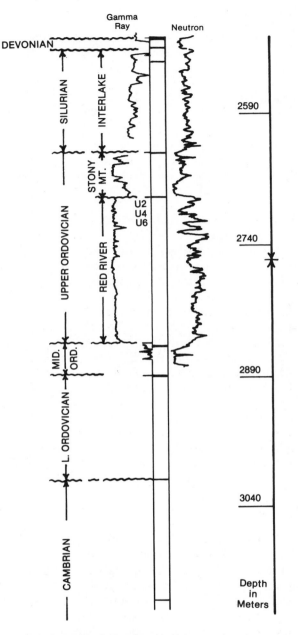

Figure 3-2 Lower Paleozoic carbonate formations in the Williston Basin together with type radioactive log (after Roehl, 1967, p. 1982).

Example 1 105

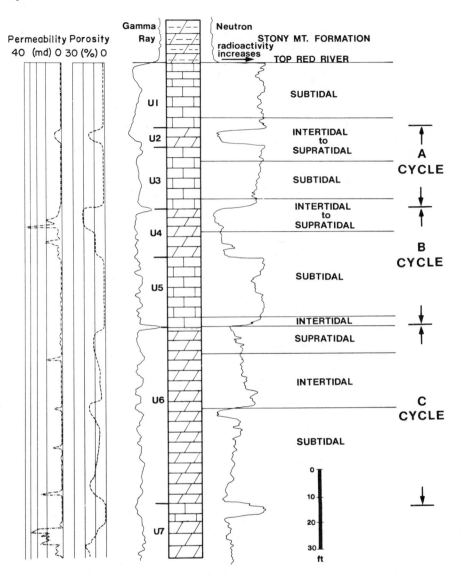

Figure 3-3 A columnar section through the upper Red River Formation showing radio-activity logs, depositional environments, porosity, and permeability (from Ruzyla and Friedman, 1980, p. 632).

In the intertidal and subtidal sediments that underlie the supratidal deposits, the degree of dolomitization decreases downward (Fig. 3-4). In these underlying units, dolomite has selectively replaced skeletal fragments and matrix, and in some parts of the sequence a gradation is evident from dolomite into unaltered limestone.

The vertical distribution of dolomite in the depositional cycles of the upper Red River Formation is in keeping with a model of downward-moving hypersaline waters

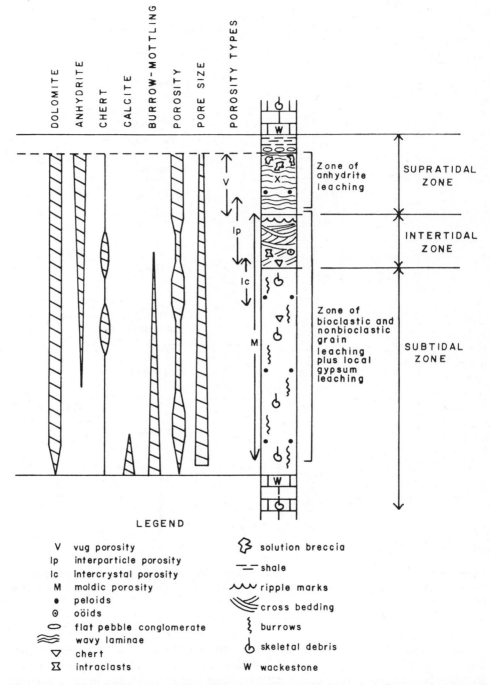

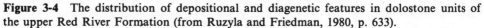

Figure 3-4 The distribution of depositional and diagenetic features in dolostone units of the upper Red River Formation (from Ruzyla and Friedman, 1980, p. 633).

Example 2 107

that originate in the supratidal zone. This movement is accompanied by progressive dolomitization of the underlying intertidal and subtidal deposits.

Distribution of Porosity

Porosity in the Red River Formation is restricted to the dolomitized units. Four main kinds of porosity occur: interparticle, intercrystal, vug, and moldic (Fig. 3-4). The development of this porosity is largely controlled by dolomitization and vadose leaching although the interparticle porosity is a primary feature that has survived compaction and diagenesis.

Interparticle porosity is most common in the high-energy intertidal zone deposits. It occurs between peloids, intraclasts, oöids, and skeletal debris in rocks containing some lime-mud matrix. Interpeloidal porosity occurs locally in the supratidal deposits.

Intercrystal porosity is the result of pervasive dolomitization and occurs to different degrees in the supratidal through subtidal deposits. Porosities of up to 20% and permeabilities of over 100 md occur where intercrystal porosity is well developed. In some horizons secondary anhydrite and silica cement have infilled the intercrystalline pore spaces.

Vug porosity and moldic porosity occur throughout the sequence as a result of leaching of evaporite minerals and carbonate particles, either before or after dolomitization.

Conclusions

Oil production from the Red River Formation is from three dolostone units interbedded with nonproductive limestones. The development of the dolostones is a function of the depositional environment, specifically the occurrence of hypersaline and freshwater conditions in the supratidal zone during deposition of parts of the sequence.

Hypersalinity led to the precipitation of evaporite minerals and syngenetic dolomitization of supratidal deposits. The downward percolation of hypersaline waters also dolomitized parts of the underlying intertidal and subtidal deposits and led to the precipitation of secondary evaporite minerals in these sediments. A later period of vadose leaching, which also originated in the supratidal environment, led to the formation of vug and moldic porosity as a result of the dissolution of evaporite minerals and carbonate particles.

The Red River Formation provides an example of a carbonate reservoir whose position is controlled by both the depositional environments and the early diagenetic processes associated with that environment.

EXAMPLE 2. A DOLOMITIC PERITIDAL RESERVOIR FROM THE JURASSIC OF AQUITAINE, SOUTHWEST FRANCE

J. Bouroullec and R. Deloffre

This study attempts to define the limits of a dolomitic reservoir unit within a given geographical area (Fig. 3-5). Data are provided from five wells arranged along an east-west profile. The subdivision of the rock units along this profile was based on

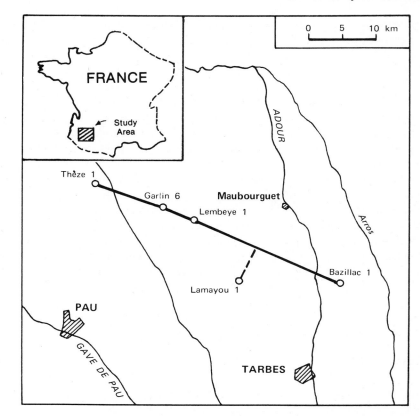

Figure 3-5 Index map of the Aquitaine Basin showing the wells studied and the position of the cross section in Figures 3-6, 3-8, and 3-9.

information from gamma-raylogs, soniclogs, and laterologs; five major horizons were delineated (Table 3-1).

Microfacies were studied in each well and the depositional environments for the sequence interpreted. The sequence involves the evolution from a lime-mud shelf environment into dolomitic tidal flats with supratidal evaporites. From the distribution of these facies in both a horizontal and vertical direction the size and distribution of the dolomitic reservoir facies can be predicted.

Data Provided by the Wells

Examination of log data from the five wells shown in Figure 3-6 defines five major horizons above the Lias (Table 3-1). Each of the three lower horizons passes to the east from limestone into dolomite, as illustrated in Figure 3-6 and Table 3-1.

A study of microfacies explains the paleogeographic significance of the different lithologies (Fig. 3-7). The characteristics used in this study include lithological, biological, dynamic, and diagenetic aspects of the different environments. These characteristics make it possible to distinguish a succession of environments related to a theoretical paleogeographic profile (Fig. 3-7) and to the different horizons or formations delineated by the logs. The paleogeographic profile shows a range of

Example 2 109

Table 3-1 Relationship between Log Horizons, Lithological Subdivisions, and Stratigraphy of the Dogger and Malm, Aquitaine, Southwest France

Main Log Horizons			Major Lithological Components		Stratigraphical Subdivisions
W		E			
Mano Formation		5	Dolomite and anhydrite		Upper Kimmeridgian to Portlandian
Lons Formation		4	Limestone		Lower Kimmeridgian
	OS 1	3	Limestone	Dolomite	
Ammonitic Marl	OS 2	2	Limestone Marls	Dolomite	Oxfordian
Limestone with Microfilaments	OS 3	1	Limestone	Dolomite	Dogger

(Ossun Formation)

depositional settings (numbered 1 to 17 in Fig. 3-7), which increase in energy level from the subtidal to tidal flat environment, related to the decrease in average water depth. Most of the low-energy limestone deposits are interpreted to occur in the subtidal area of the shelf, whereas the lower two dolomitic formations are believed to occur in the fairly high-energy tidal-flat environment. The upper dolomitic formation is interpreted as an evaporitic supratidal deposit.

Figure 3-8 shows the vertical distribution of lithologies and textural components in each of the wells studied. By comparing their distribution in neighboring wells a series of paleogeographic profiles can be drawn (Fig. 3-9) and the main sedimentary environments defined in terms of the observed features in the deposits.

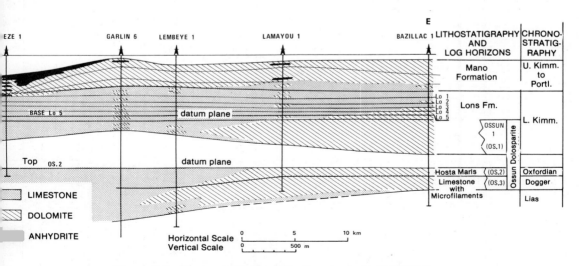

Figure 3-6 Cross section along the line shown in Figure 3-5.

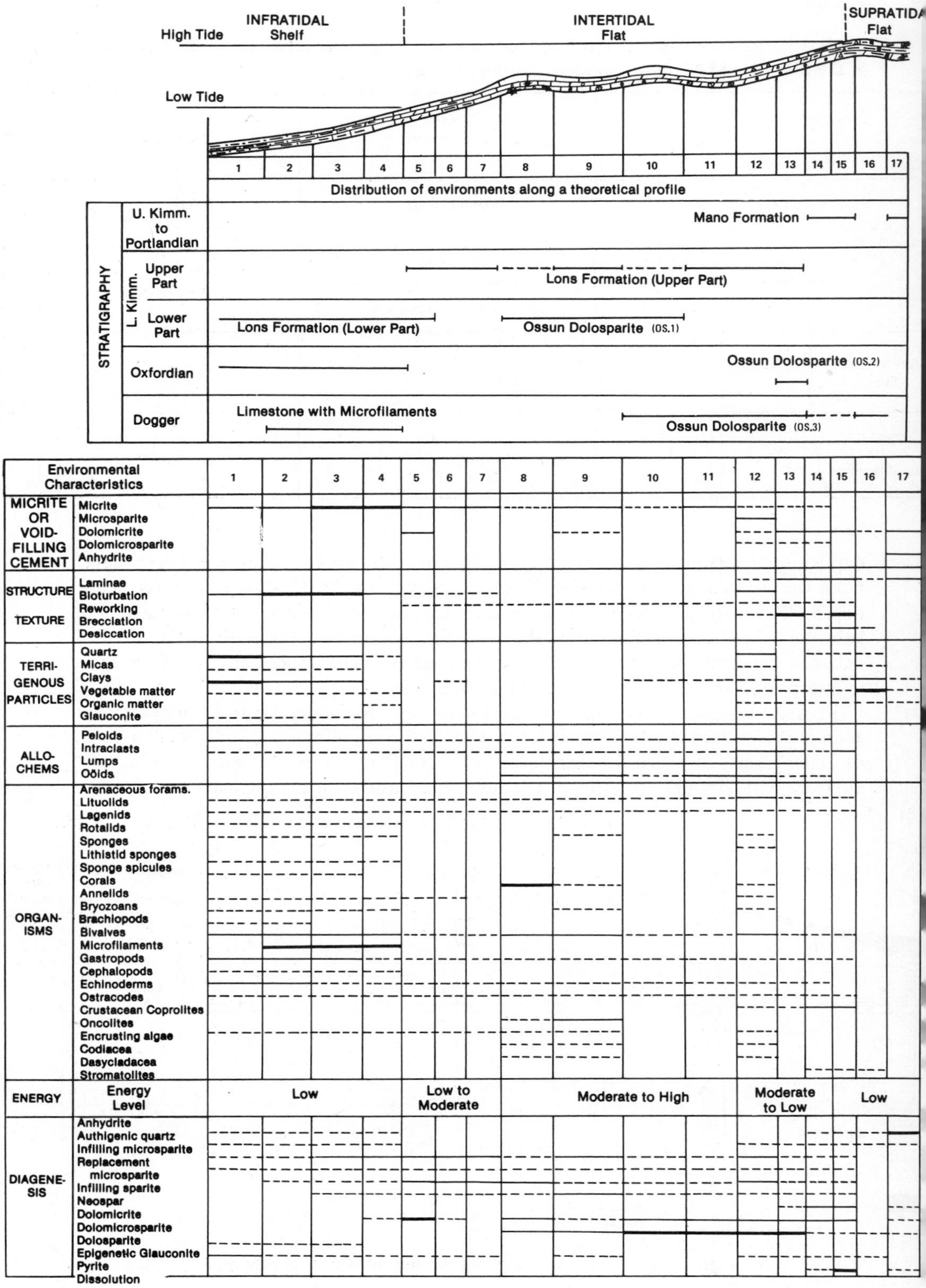

INFRATIDAL
Shelf

INTERTIDAL
Flat

SUPRATIDA
Flat

Low Tide

| 1 | 2 | 3 | 4 | 5 | 6 | 7 | 8 | 9 | 10 | 11 | 12 | 13 | 14 | 15 | 16 | 17 |

Distribution of environments along a theoretical profile

STRATIGRAPHY

U. Kimm. to Portlandian		Mano Formation
L. Kimm. — Upper Part		Lons Formation (Upper Part)
L. Kimm. — Lower Part	Lons Formation (Lower Part)	Ossun Dolosparite (OS.1)
Oxfordian		Ossun Dolosparite (OS.2)
Dogger	Limestone with Microfilaments	Ossun Dolosparite (OS.3)

Environmental Characteristics		1	2	3	4	5	6	7	8	9	10	11	12	13	14	15	16	17

MICRITE OR VOID-FILLING CEMENT
- Micrite
- Microsparite
- Dolomicrite
- Dolomicrosparite
- Anhydrite

STRUCTURE / TEXTURE
- Laminae
- Bioturbation
- Reworking
- Brecciation
- Desiccation

TERRIGENOUS PARTICLES
- Quartz
- Micas
- Clays
- Vegetable matter
- Organic matter
- Glauconite

ALLOCHEMS
- Peloids
- Intraclasts
- Lumps
- Oölds

ORGANISMS
- Arenaceous forams.
- Lituolids
- Lagenids
- Rotalids
- Sponges
- Lithistid sponges
- Sponge spicules
- Corals
- Annelids
- Bryozoans
- Brachiopods
- Bivalves
- Microfilaments
- Gastropods
- Cephalopods
- Echinoderms
- Ostracodes
- Crustacean Coprolites
- Oncolites
- Encrusting algae
- Codiacea
- Dasycladacea
- Stromatolites

ENERGY — Energy Level: Low | Low to Moderate | Moderate to High | Moderate to Low | Low

DIAGENESIS
- Anhydrite
- Authigenic quartz
- Infilling microsparite
- Replacement microsparite
- Infilling sparite
- Neospar
- Dolomicrite
- Dolomicrosparite
- Dolosparite
- Epigenetic Glauconite
- Pyrite
- Dissolution

Figure 3-7 Distribution of microfacies in the Jurassic of the Aquitaine Basin, southwestern France.

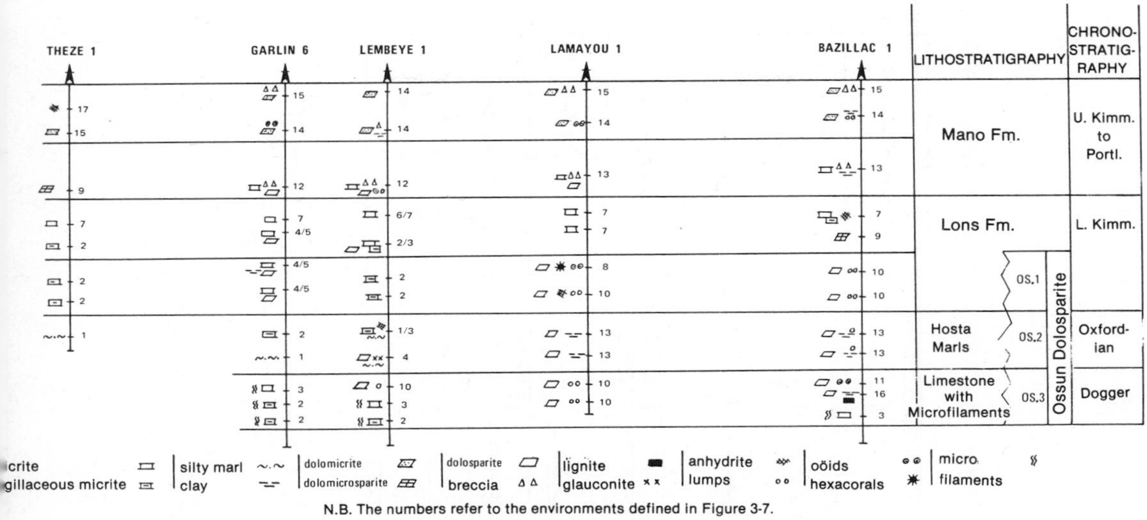

Example 2 111

THEZE 1	GARLIN 6	LEMBEYE 1	LAMAYOU 1	BAZILLAC 1	LITHOSTRATIGRAPHY	CHRONO-STRATIGRAPHY
17	15	14	15	15	Mano Fm.	U. Kimm. to Portl.
15	14	14	14	14		
9	12	12	13	13		
7	7	6/7	7	7	Lons Fm.	L. Kimm.
2	4/5	2/3	7	9		
2	4/5	2	8	10	OS.1	
2	4/5	2	10	10		
1	2	1/3	13	13	Hosta Marls — OS.2	Oxfordian
	1	4	13	13		
	3	10	10	11	Limestone with Microfilaments — OS.3	Dogger
	2	3	10	16		
	2	2		3		

(Ossun Dolosparite: OS.1, OS.2, OS.3)

Legend:

micrite	silty marl	dolomicrite	dolosparite	lignite	anhydrite	ooids	micro. filaments
argillaceous micrite	clay	dolomicrosparite	breccia	glauconite	lumps	hexacorals	filaments

N.B. The numbers refer to the environments defined in Figure 3-7.

Figure 3-8 Cross section along the line in Figure 3-5, showing data obtained from both log and sample studies.

The subtidal environment is dominated by micrite, argillaceous micrite, and silty, ammonite-bearing marls. The basal unit of the sequence is a micrite containing microfilaments (filamentous particles of uncertain origin, possibly algal) that forms a distinct basal zone and extends upward in places into the Lons Formation.

The intertidal zone consists of dolosparites, and it is these deposits that are the reservoir objectives. Deposition occurred over a well-developed tidal flat with gravelly and oölitic buildups indicating periods of high energy, especially in the outer intertidal zone, where frame-building organisms occur. Intertidal deposits that formed on local high shelf ridges are found interfingered with shelf deposits in the well Garlin 6.

The supratidal deposits are dolomicrites, anhydrites, and brecciated dolomicrites formed by the dissolution of interbedded evaporites. These deposits constitute most of the Mano Formation. An additional facies, a dolomite with lignitic seams, occurs locally in the Bazillac 1 well at the OS 3 horizon.

Conclusions

Within the five horizons differentiated in this sequence the evolution of environments in an east-west section has been reconstructed.

The change from limestone to dolomite corresponds to the passage from shelf into tidal-flat environments. During the Middle to Late Jurassic there was progressive displacement of the intertidal dolomitic facies toward the east. This displacement represents the movement of the Jurassic shoreline in time and space. Figure 3-10 is a plan representation of the movement of the intertidal zone during the Middle and Late Jurassic. The arrows in the diagram show the sense of the movement compared to the position of the shoreline in the preceding time period.

In Figure 3-11 the potential reservoirs together with the maximum crystal size of the dolomite are shown for the OS 1 tidal-flat deposits. The porosity in these dolomites is mainly intercrystalline and is directly related to crystal size. Figure 3-11 shows an

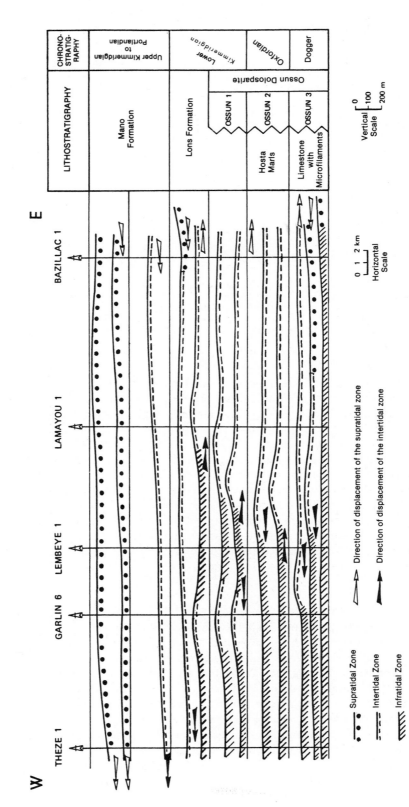

Figure 3-9 Distribution of interpreted peritidal zones in wells along line of section shown in Figure 3-5.

Example 3   113

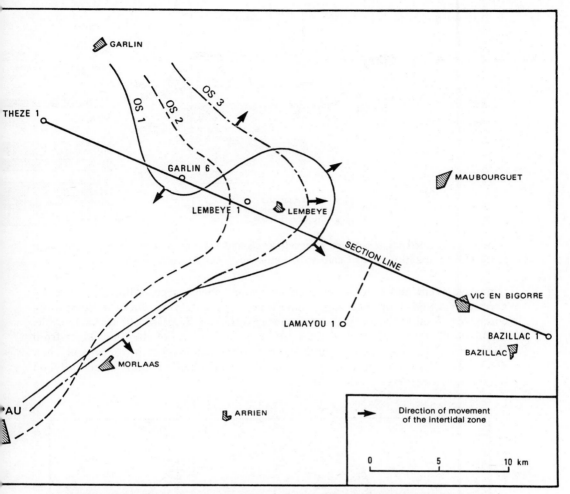

Figure 3-10   Movement of intertidal zone during the Middle and Late Jurassic.

increase in the number and thickness of reservoir horizons from west to east. The maximum crystal size of the dolomite shows a general increase in the reservoir horizons, particularly in the thickest horizons in the Lamayou 1 and Bazillac 1 wells. Outside the reservoir horizons the maximum crystal size decreases. The crystal size-porosity relationship implies that porosity improves with increasing development of dolomite in the interpreted tidal flats, thus creating favorable reservoir horizons in rocks deposited in this environment.

EXAMPLE 3.   HIGH-ENERGY SHALLOW MARINE RESERVOIRS IN THE SMACKOVER FORMATION (JURASSIC), GULF COAST, U.S.A.

The Smackover Formation and overlying Buckner Formation of Jurassic age occur in an arcuate trend in the subsurface marginal to the Gulf Basin of the southern United States (Fig. 3-12). The Smackover Formation was deposited during a transgressive-

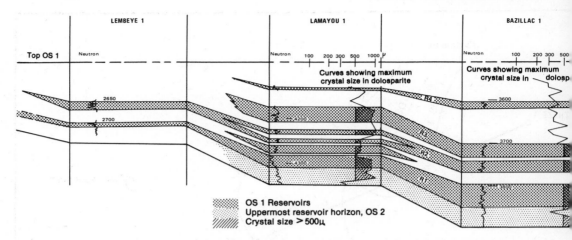

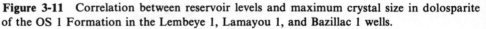

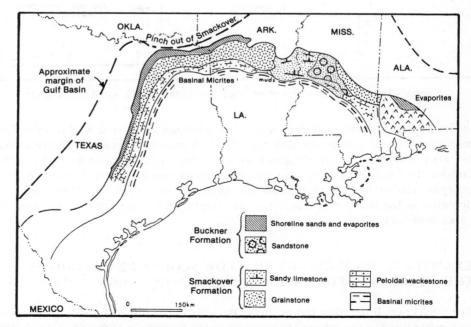

Figure 3-11 Correlation between reservoir levels and maximum crystal size in dolosparite of the OS 1 Formation in the Lembeye 1, Lamayou 1, and Bazillac 1 wells.

regressive event and consists of shelf dolomites, shelf-margin oölitic and peloidal grainstones, and slope and basin limestones. The overlying Buckner Formation, composed of anhydrite and red beds, represents the final regressive phase of the cycle.

Oil production is limited to the upper regressive phase of the Smackover from reservoir rocks that include high-energy oölitic-peloidal grainstones and shelf dolomites. The location of the reservoirs is controlled by both structural and stratigraphic features (Collins, 1980).

Figure 3-12 Distribution of lithologies during deposition of the upper Smackover Formation (after Bishop, 1968; Wilson, 1975; Ottman et al., 1976).

Example 3   115

Geological Setting

The oölitic-peloidal grainstones of the upper Smackover Formation are restricted to a belt approximately 100 m thick and 8 to 30 km wide which is semiparallel to the margin of the Gulf Basin (Fig. 3-12). These high-energy grainstones are interpreted as shelf-edge deposits whose position is controlled by a tectonic hinge that forms the shelf margin (Wilson, 1975). The shelf-edge oölites and grainstones pass basinward through a peloidal bioclastic wackestone slope facies into a thick sequence of basinal deposits mostly assigned to the laterally equivalent Bossier Formation. The Smackover itself is largely a basin-margin formation which loses its identity when traced basinward (Dickinson, 1968; Becher and Moore, 1976). Traced landward the Smackover passes into the evaporites and red beds of the Buckner Formation (Fig. 3-13).

The Smackover Formation has a thickness of 250 to 400 m, but it is only the upper 100 m or less that are of interest as potential reservoir strata. Porosity in this interval has diverse origins and may be either primary, early, or late diagenetic.

Vertical Sequence

The Smackover Formation overlies clastic sandstones, conglomerates, and shales of the Norphet Formation which have been interpreted as including fluvial, dune, deltaic, and shore-face deposits (Sigsby, 1976). The basal Smackover lithologies include laminated mudstones, micrites, and dark brown silty to argillaceous micrites containing peloids and foraminifera (Dickinson, 1968). In basin-margin settings in Alabama, the basal Smackover contains laminated structures interpreted as intertidal algal-mat deposits which pass upward with marine transgression into open marine peloidal and skeletal micrites (Ottmann et al., 1976; Sigsby, 1976).

The middle section of the Smackover constitutes over two-thirds of the entire formation. It consists of dense brown limestones and dolostones with scattered peloids and oöids (Dickinson, 1968). This interval in some sequences marks the period of maximum transgression and the beginning of the regressive phase of the formation. In contrast to the upper Smackover, which has been extensively studied in a number of

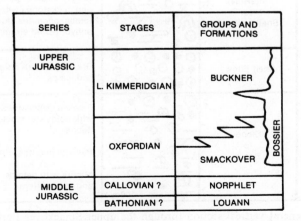

Figure 3-13 Stratigraphy of the Smackover Formation and adjacent rock units (after Dickinson, 1968; Ottmann et al., 1976).

areas because of its petroleum potential, little is known about the lower and middle part of the formation. The lower and middle Smackover is generally described as fine-grained, dark in color, and micritic, but interpretations of the environment in which it was deposited range from intertidal to deep basin (Dickinson, 1969; Ottmann et al., 1976).

The upper Smackover represents the regressive phase of deposition and the sequence is interpreted as passing upward from slope deposits to shelf sediments. The general vertical succession shows a gradation from peloidal micrites and dolomite into packstone-wackestones containing scattered oöids, peloids, and bioclastic debris. With increasing abundance of grains these pass upward into grainstones composed of oöids, peloids, and lumps with very minor bioclastic debris. This vertical gradation is interpreted as a general shallowing sequence from the shelf-slope to the high-energy setting of the shelf edge. The lack of fossil debris in the grainstone facies is interpreted as evidence of increasing salinity (Bishop, 1968).

The oölitic and peloidal grainstones pass upward into wackestones and packstones containing peloids, öoids, lumps, algal oncolites, and some quartz sand. These may become shaley and contain dolomite, anhydrite, and cryptalgal laminites. They represent low-energy shelf to intertidal deposits which grade upward into the evaporites and red beds of the Buckner Formation (Fig. 3-14).

The vertical section described here is a general section only and illustrates the regressive nature of the formation. There is much regional variation within the upper Smackover in both the vertical sequences of lithologies that are developed and in the

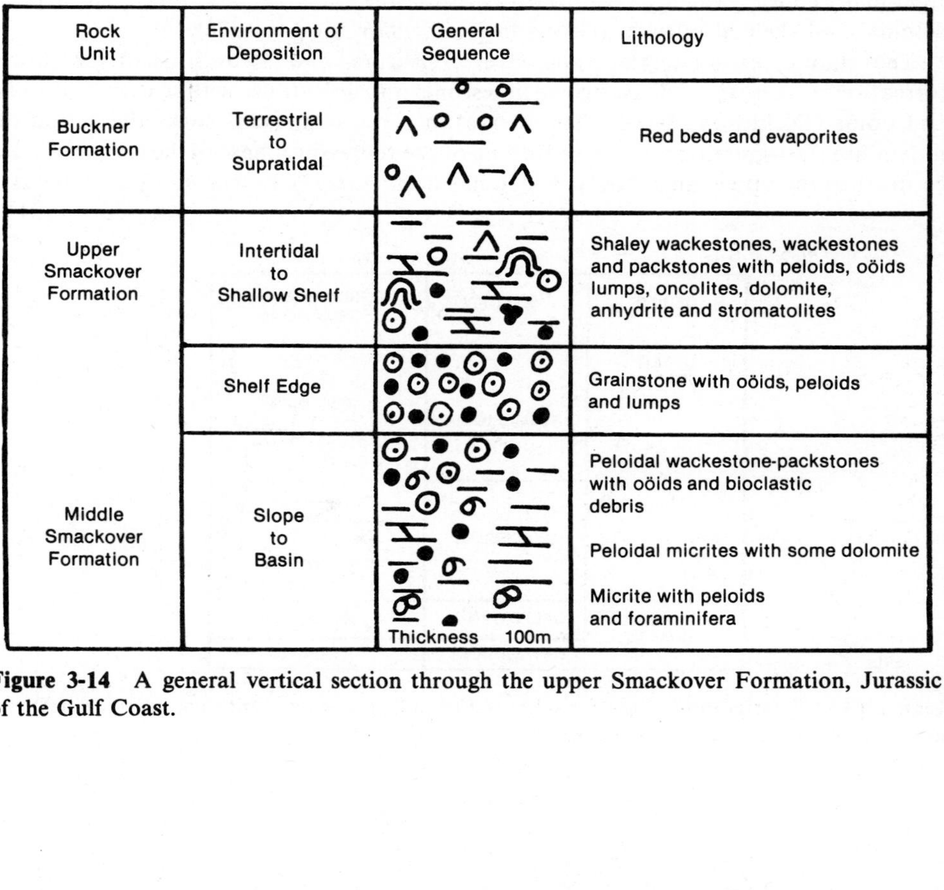

Rock Unit	Environment of Deposition	General Sequence	Lithology
Buckner Formation	Terrestrial to Supratidal		Red beds and evaporites
Upper Smackover Formation	Intertidal to Shallow Shelf		Shaley wackestones, wackestones and packstones with peloids, oöids lumps, oncolites, dolomite, anhydrite and stromatolites
	Shelf Edge		Grainstone with oöids, peloids and lumps
Middle Smackover Formation	Slope to Basin		Peloidal wackestone-packstones with oöids and bioclastic debris
			Peloidal micrites with some dolomite
			Micrite with peloids and foraminifera
		Thickness 100m	

Figure 3-14 A general vertical section through the upper Smackover Formation, Jurassic of the Gulf Coast.

Example 3 117

kinds of particles that make up these lithologies. These variations are in part the result of minor regressive-transgressive episodes within the general regressive megasequence and in part the result of local topographical variations on the seafloor during deposition of the upper Smackover.

Horizontal Development of Lithologies

Figure 3-15 is a depositional model for the upper Smackover which shows the vertical and lateral distribution of facies and environments. Deposition during upper Smackover time took place across a narrow shelf, where environments graded from terrestrial to shelf edge. Grainstone bars developed at the shelf edge and acted as partial barrier to water movements on the inner shelf, where hypersaline conditions and the precipitation of evaporites occurred. The grainstone bars and associated slope deposits were progradational across the basinal micrites; thus the overall vertical sequence shows a shallowing-upward or regressive nature.

Smackover Reservoirs

Reservoirs are mainly restricted to packstone and grainstone lithologies in the upper Smackover. Variations in amount and kind of porosity within the Smackover are apparently the result of differing diagenetic histories. Porosity in the reservoirs is of three main kinds: (1) primary interparticle porosity, (2) leached moldic porosity, and (3) intercrystal porosity within dolomites. In Arkansas, Louisiana, and eastern Mississippi the best reservoirs have incompletely cemented primary interparticle porosity. The updip Smackover close to the paleoshoreline contains leached moldic porosity developed as a result of freshwater flushing. In eastern Texas, Alabama, and Florida dolomites with intercrystal porosity occur in addition to leached moldic porosity (Stoudt, 1979).

Porosity in the Walker Creek field, Arkansas, is postulated as depending on early postdepositional structural movements accompanied by subaerial exposure (Fig. 3-16). Diagenesis during these periods produced some cementation, dolomitization,

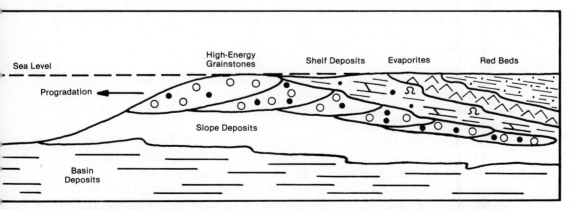

Figure 3-15 Depositional model for the upper Smackover showing the lateral and vertical distribution of facies and environments (modified after Reading, 1978, p. 292).

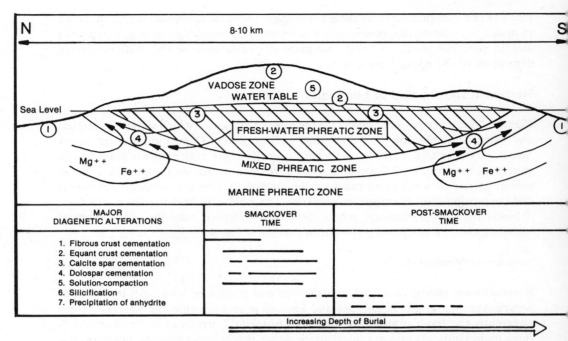

Figure 3-16 Schematic cross section through the Walker Creek field, Arkansas, during Smackover time. A freshwater lens is present within the topographic high. Dolomite formed in the mixing zone of freshwater and seawater. The distribution of diagenetic features is shown by the corresponding numbers on the cross section, and the timing of the alterations is indicated by the horizontal lines beneath the section (after Becher and Moore, 1976).

and solution in the meteoric phreatic zone, while primary porosity was preserved in the overlying vadose zone (Becher and Moore, 1976).

In general, porosity destruction in grainstones within the Smackover is the result of cementation by calcite spar, void-filling dolomite and anhydrite, compaction, and pressure solution. Formation of porosity is the result of replacement dolomitization and freshwater leaching of allochems, especially oöids.

Conclusions

The Smackover Formation was deposited in the Gulf Coast Basin of the southern United States during a major Jurassic transgressive-regressive event. The formation is composed of shelf deposits that grade basinward into deep-water micrites and shoreward through evaporites and red beds.

The upper Smackover and overlying Buckner Formation form the regressive part of the cycle and are composed of slope wackestone-packstones, shelf-edge grainstones, and low-energy shelf wackestones, dolomites, and evaporites. Reservoirs are essentially confined to the grainstone-packstone lithologies of the upper Smackover and produce from primary interparticle porosity or secondary moldic porosity. Dolomitized shelf sediments, peritidal dolomites, and dolomitized stromatolites immediately underlying the Buckner Formation are also producing, and these lithologies have intercrystal porosity.

Example 4   119

EXAMPLE 4. CARBONATE SHELF RESERVOIRS: THE MIDDLE JURASSIC OF THE PARIS BASIN, FRANCE

R. Cussey, E. Grosdidier, L. Sulpice, and P. Umbach

The Middle Jurassic of the Paris Basin has been the object of numerous syntheses. This study is original in that it combines information obtained from sedimentary analysis and well logging at many levels of detail (Table 3-2). Data were used from earlier

Table 3-2 Relationship between Sedimentological and Well-Logging Studies Dealing with the Synthesis of the Dogger (Middle Jurassic) of the Paris Basin

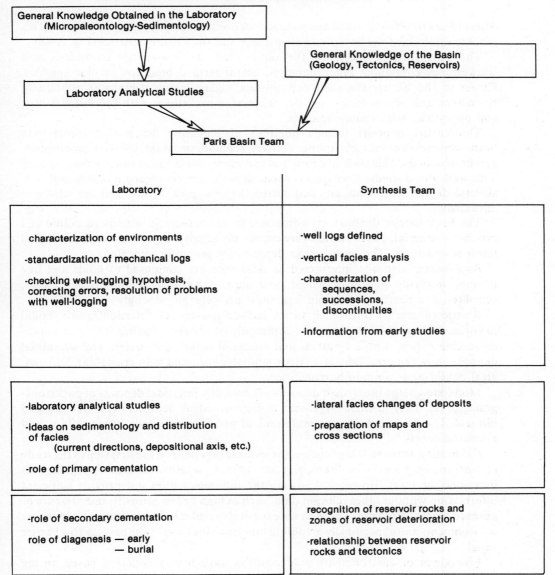

studies including an additional 100 boreholes, and previous paleogeographic and environmental interpretations were taken into account. Among the most recent of these is the study of Purser and Loreau (1972) (Fig. 3-17).

Wells previously studied were reinterpreted in terms of their depositional environment. Facies were established using logs and interpreted together with depositional environments and their distribution in time and space. These methods were part of a rapid survey whose aim was to supply supportive sedimentological data. Frequent interdisciplinary contact throughout the study allowed the development and modification of ideas and significantly enhanced the outcome of the project.

Analysis and Interpretation of the Deposits

Microfacies were first defined and assigned to specific depositional environments. This quickly established the environments present and their characteristics (Fig. 3-18).

The open marine facies of the outer shelf are low-energy mudstones and wackestones containing microfilaments, echinoderm debris, and sponge spicules. Closer to the barrier the shelf deposits are higher-energy packstones containing bryozoans and echinoderms, oölitic wackestone-packstones with bioclastic debris, and packstones with sponge spicules.

The barrier deposits include buildups of bioclastic debris that incorporate bryozoans, small coral reefs, sponge clusters, echinoderm banks, bivalves, gastropods, and brachiopods. This skeletal debris was commonly reworked *in situ* and is associated with well-sorted oöids. Contiguous shoal deposits are composed of oöids and lack skeletal debris. The oöids are well sorted, have a thick cortex, and are relatively uncemented.

The back-barrier deposits are composed of reworked and winnowed oölitic and bioclastic material. These particles are commonly micritized with loss of their original internal structures. The back-barrier deposits are generally cemented.

Back-barrier deposits interpreted as tidal flats are composed of oöids and fine peloids, normally poorly cemented, and more commonly of intraclasts, peloids, oncolites, and benthonic foraminifera; these are generally strongly cemented.

Restricted marine inner-shelf facies include low-energy intertidal and subtidal deposits. The subtidal deposits are commonly of micrite together with fine microcrystalline debris, whole bivalves, and scattered algae (*Cayeuxia*). The supratidal deposits show fenestral fabrics, desiccation polygons, mud with crustacean burrows, algal debris, and common burrow traces.

Moderate-energy inner-shelf deposits are basically intertidal deposits of packstone-grainstones. They include algal-mat sediments, algal deposits with peloids and intraclasts, algal oncolites, accumulations of pellets, gastropods, foraminifera, and bioclastic debris.

Calibration of the well logs against the sedimentary facies is the first step in the study of instrumental logs. The lithologies are defined, whether limestone, argillaceous limestone, or marl. However, knowing the lithologies does not provide sufficient information without other relevant data, such as high or low porosity and presence or absence of clay. High porosity with absence of clay and micrite indicates a high-energy environment, more specifically an oölitic limestone that marks the passage to a barrier shoal.

Five kinds of environments of deposition have been recognized based on log

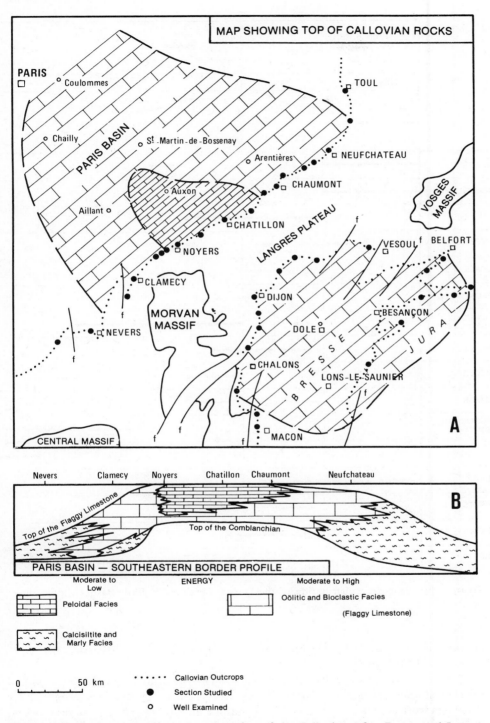

Figure 3-17 Paleogeographic map and section of the Callovian (after Purser and Loreau, 1972).

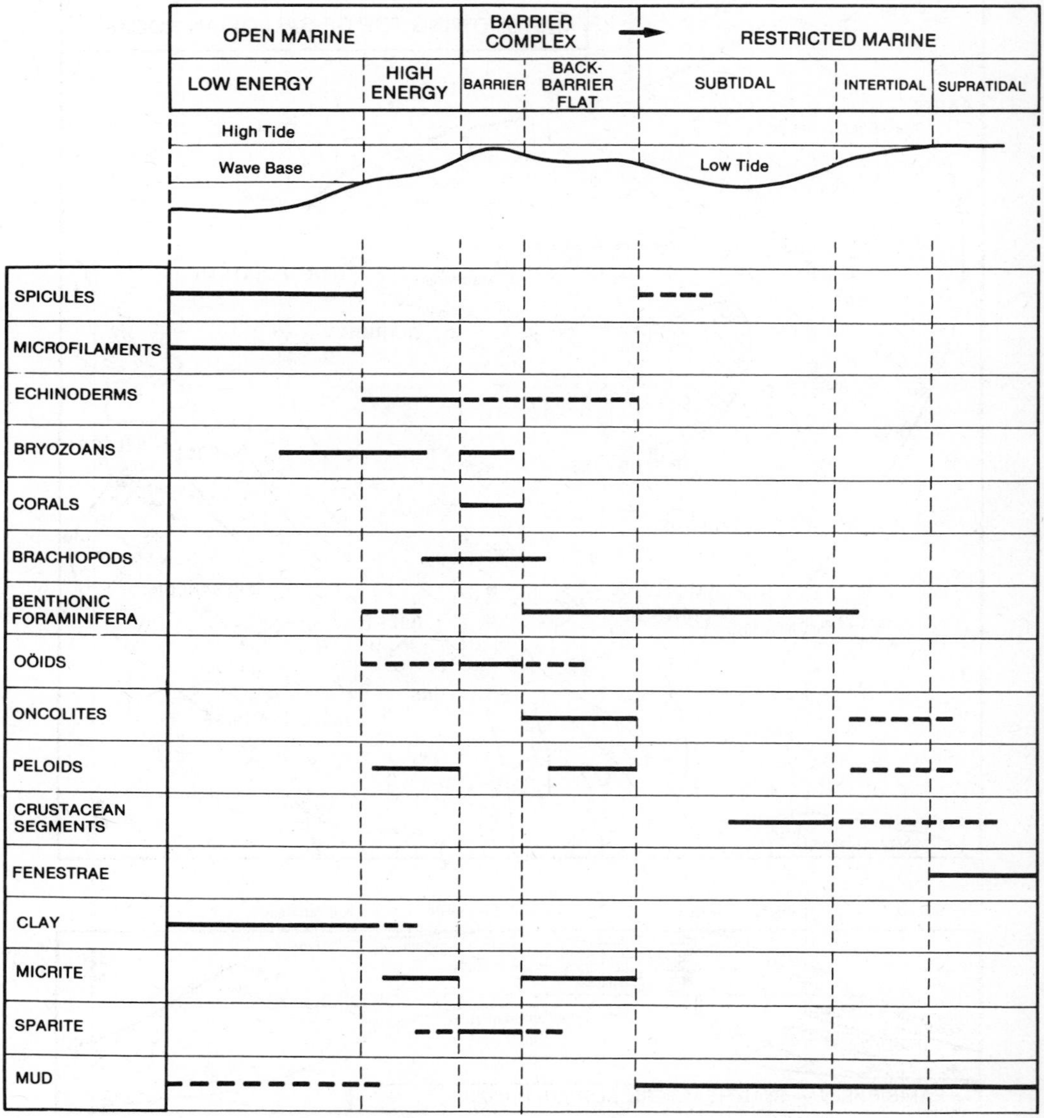

OPEN MARINE		BARRIER COMPLEX →		RESTRICTED MARINE		
LOW ENERGY	HIGH ENERGY	BARRIER	BACK-BARRIER FLAT	SUBTIDAL	INTERTIDAL	SUPRATIDAL

High Tide
Wave Base — Low Tide

SPICULES

MICROFILAMENTS

ECHINODERMS

BRYOZOANS

CORALS

BRACHIOPODS

BENTHONIC FORAMINIFERA

OÖIDS

ONCOLITES

PELOIDS

CRUSTACEAN SEGMENTS

FENESTRAE

CLAY

MICRITE

SPARITE

MUD

Figure 3-18 Schematic distribution of the Dogger facies of the Paris Basin (Middle Jurassic).

characteristics (Fig. 3-19): (1) open marine, (2) fore-barrier, (3) barrier, (4) back-barrier, and (5) lagoon.

The open marine deposits are composed of marls, argillaceous limestones, and microcrystalline limestones. The radioactivity measured by the logs progressively decreases from the marls into the microcrystalline limestones. The SP curve shows no deflections. The marls show very low resistivity, which increases regularly with the amount of carbonate present. The neutron-hydrogen index decreases from the marls into the limestones.

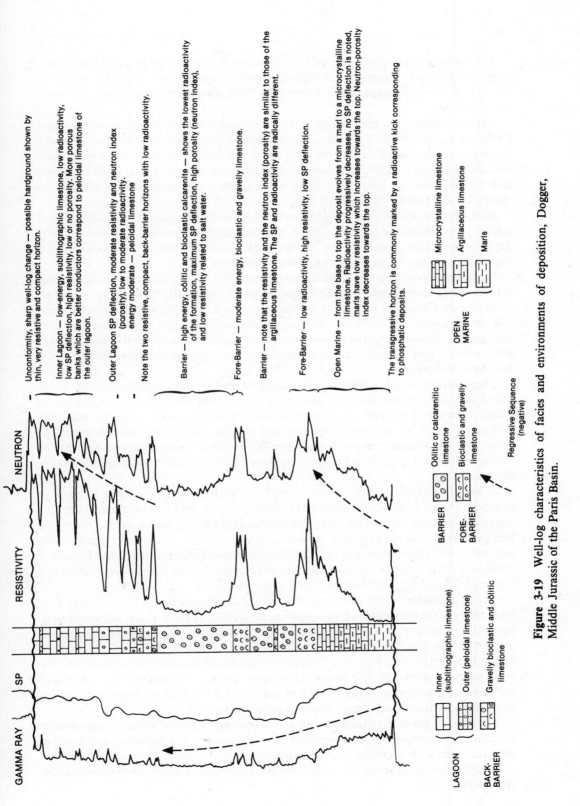

Unconformity, sharp well-log change — possible hardground shown by thin, very resistive and compact horizon.

Inner Lagoon — low-energy, sublithographic limestone, low radioactivity, low SP deflection, high resistivity, low or no porosity. More porous banks which are better conductors correspond to peloidal limestone of the outer lagoon.

Outer Lagoon SP deflection, moderate resistivity and neutron index (porosity), low to moderate radioactivity. energy moderate — peloidal limestone

Note the two resistive, compact, back-barrier horizons with low radioactivity.

Barrier — high energy, oölitic and bioclastic calcarenite — shows the lowest radioactivity of the formation, maximum SP deflection, high porosity (neutron index), and low resistivity related to salt water.

Fore-Barrier — moderate energy, bioclastic and gravelly limestone.

Barrier — note that the resistivity and the neutron index (porosity) are similar to those of the argillaceous limestone. The SP and radioactivity are radically different.

Fore-Barrier — low radioactivity, high resistivity, low SP deflection.

Open Marine — from the base to top the deposit evolves from a marl to a microcrystalline limestone. Radioactivity progressively decreases, no SP deflection is noted, marls have low resistivity which increases towards the top. Neutron-porosity index decreases towards the top.

The transgressive horizon is commonly marked by a radioactive kick corresponding to phosphatic deposits.

Figure 3-19 Well-log characteristics of facies and environments of deposition, Dogger, Middle Jurassic of the Paris Basin.

LAGOON — Inner (sublithographic limestone)
Outer (peloidal limestone)

BACK-BARRIER — Gravelly bioclastic and oölitic limestone

BARRIER — Oölitic or calcarenitic limestone

FORE-BARRIER — Bioclastic and gravelly limestone

Regressive Sequence (negative)

OPEN MARINE — Microcrystalline limestone
Argillaceous limestone
Marls

GAMMA RAY   SP   RESISTIVITY   NEUTRON

The fore-barrier deposits are moderate-energy bioclastic and peloidal limestones. They are slightly radioactive with high resistivity, which is higher than that shown by the open marine microcrystalline limestones. They have no porosity and show slight SP deflection.

Barrier deposits are high-energy deposits composed of oölitic limestone (Dogger Oölite) and calcarenites. They show minimal radioactivity, less than the other deposits, and have maximum SP deflection. They show low resistivity because the pores have been filled with salt water. They have a high neutron-hydrogen index. If the resistivity and neutron index are the same as those of the argillaceous limestones the difference is shown by the gamma-ray and SP logs.

Back-barrier deposits are moderate-energy peloidal, bioclastic, and oölitic limestones. The oölitic limestones have a cryptocrystalline cement. The back-barrier deposits show high resistivity, low radioactivity, and an SP deflection that is less then that shown by the barrier deposits. Their porosity is low.

Lagoonal deposits can be divided into inner and outer. The outer lagoonal deposits are moderate-energy peloidal limestones that form banks. They have moderate neutron-index porosity and moderate resistivity and show strong SP deflection and low to moderate radioactivity. The inner lagoonal deposits are low-energy sublithographic limestones. They have low or no porosity, no SP deflection, average radioactivity, and very strong resistivity, equal to or greater than that shown by the fore-barrier deposits.

Characterization of a deposit taken in isolation was often impossible and it was found necessary to consider each deposit in terms of the evolution of the sequence in which it occurs. The breaks and changes in sedimentation are marked by sharp log deflection, which allows easy correlation and interpretation of the facies present. The fore- and back-barrier deposits have similar well-log characteristics, but can be distinguished in a regressive sequence by their position relative to the oölitic barrier deposits.

In addition to the calibration of the well logs with the sedimentary facies, a large number of wells (276) were logged and interpreted in terms of vertical sequence. This approach included consideration of the defined sedimentary facies, the facies defined from well logs, and the expected vertical succession of deposits. From these analyses vertical profiles were drawn (Fig. 3-20). It can be seen from these that the Dogger is composed of a number of coarsening-upward cycles within a large-scale regressive sequence (Fig. 3-21), followed by a major transgression.

The first coarsening-upward sequence (Fig. 3-20, No. 1) begins in the Upper Lias and results in the deposition of open marine bioclastic deposits. A second coarsening-upward sequence (Fig. 3-20, No. 2) begins in the Bajocian with a rapid transgression and ends with the deposition of localized oölitic sediments. The third coarsening-upward sequence (Fig. 3-20, No. 3) occurs in the Bathonian with the deposition of a lagoonal shoal sequence. The Dogger taken as a whole is clearly regressive. The top of each coarsening-upward cycle is interpreted as being more shallow than the preceding one; open marine deposits have become uncommon.

The final stages of the Dogger begin with a transgression and a coarsening-upward sequence. Two successive sequences can be differentiated, but the latter one is rather thin and ends with a general transgression.

Lateral paleogeographic variations can be reconstructed from the vertical profiles (Fig. 3-21).

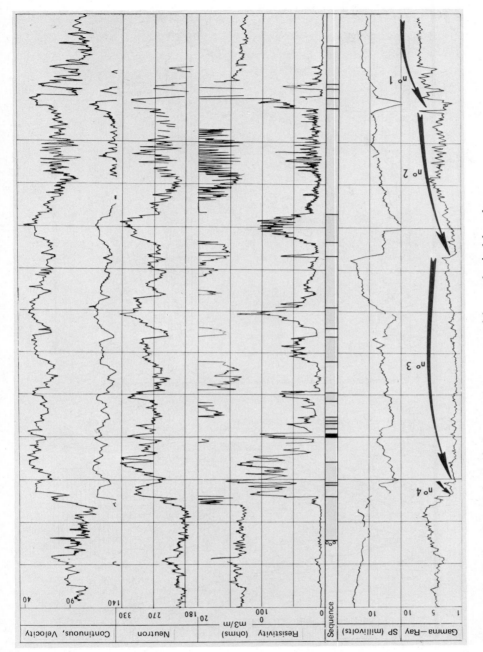

Figure 3-20 Facies changes in vertical sequence as expressed by mechanical log characteristics. See Figure 3-23 for the section key.

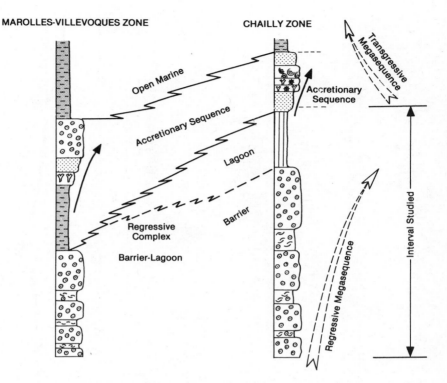

MAROLLES-VILLEVOQUES ZONE

CHAILLY ZONE

Open Marine

Accretionary Sequence

Lagoon

Barrier

Accretionary
Sequence

Transgressive
Megasequence

Regressive
Complex

Barrier-Lagoon

Regressive Megasequence

Interval Studied

(a) Vertical relationship between regressive and overlying accretionary sequence

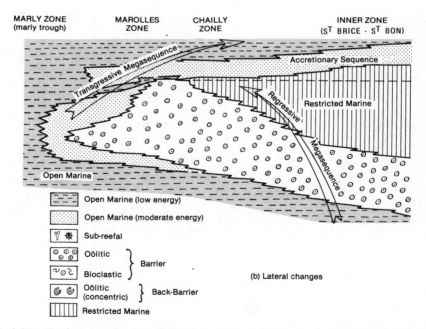

MARLY ZONE
(marly trough)

MAROLLES
ZONE

CHAILLY
ZONE

INNER ZONE
(S^T BRICE - S^T BON)

Transgressive Megasequence

Accretionary Sequence

Restricted Marine

Regressive Megasequence

Open Marine

Open Marine (low energy)

Open Marine (moderate energy)

Sub-reefal

Oölitic   ⎱
              ⎰ Barrier
Bioclastic

Oölitic   ⎱
(concentric) ⎰ Back-Barrier

Restricted Marine

(b) Lateral changes

Figure 3-21 Vertical and lateral relationships between facies in the Middle Jurassic (Dogger), in the middle of the Paris Basin.

Example 4   **127**

Paleogeographic Synthesis

The paleogeographic evolution of the Dogger falls into a specific pattern:

1 The sedimentary sequences were deposited by successive coarsening-upward cycles (accretionary sequences) (Fig. 3-21).
2 The regressive megasequence from the Lias to the Bajocian is characterized by the deposition of more restricted sediments at the top (Fig. 3-21).
3 The rapid transgressive sequence from the Bathonian onward is characterized by poorly developed accretionary sequences showing transgressive onlap (Fig. 3-21).
4 Reefal barrier deposits are absent, but vast oölitic deposits occur similar in origin to those found in the modern Bahamas. The oöid bodies developed in long banks with a consistent orientation, but irregular limits in the direction of both the open sea and the lagoons (Fig. 3-22).

The regressive evolution during the Bathonian is particularly noticeable in some sections (Fig. 3-23), where lagoonal deposits overlie oölitic and open marine deposits. This evolution through time resulted in the deposition of a large lagoon-shoal system at the end of the Bathonian. This development may have been the result of a recurrent northwest-southeast median suture that separated the stable region from the area of subsidence. Between the two areas existed an open marine trough in which marl accumulated.

In the subsiding area of the basin restricted marine deposits developed protected by the oölitic barrier. This barrier extended toward the east beyond the Morvano-Vosgian shelf as far as the Jura and then westward along the probably emergent Ardennes shelf ridge, and from Artois to the English Channel.

A transgression at the end of the Bathonian into the Callovian overlapped the previously deposited restricted marine sediments, and the distribution of deposits became very complex (Fig. 3-22). The previous distinction between oölitic and bioclastic sediments became insufficient, since there appear to be two kinds of öolite— one occurs in shoals and constitutes the reservoir facies while the other is a nonreservoir facies. Finally, at the end of the Callovian open marine marls totally occupied the basin.

Causes of Reservoir Deterioration

The reservoirs are formed solely by the oölitic and bioclastic deposits that developed as flats and shoals. Deterioration in petrophysical properties is linked with both their sedimentary and diagenetic histories.

The presence of interparticle porosity is related to the absence of diagenesis or to partial preservation as a result of moderate diagenesis. Various factors cause reservoir deterioration, which are of both primary and secondary origin and of varying importance, as will be discussed.

Reservoirs deteriorate when filled with mud or lime mud, either washed in from the open sea or from the lagoon behind. The unclogged zone therefore tends to be confined to the central belt of the shoal. This infiltration of mud controls to some extent the size of the reservoir. Other kinds of internal sediment that can be washed in and trapped between grains include microaggregates or peloids which form a micropeloidal

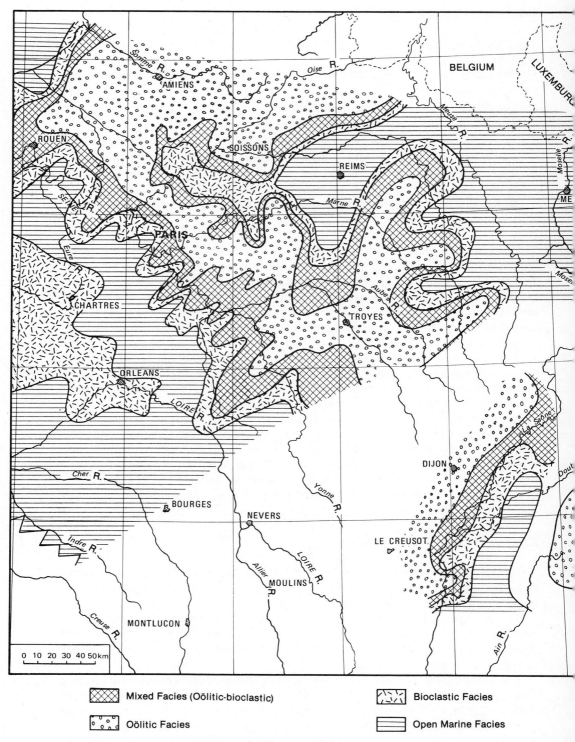

Figure 3-22 Distribution of Callovian deposits.

Example 4   129

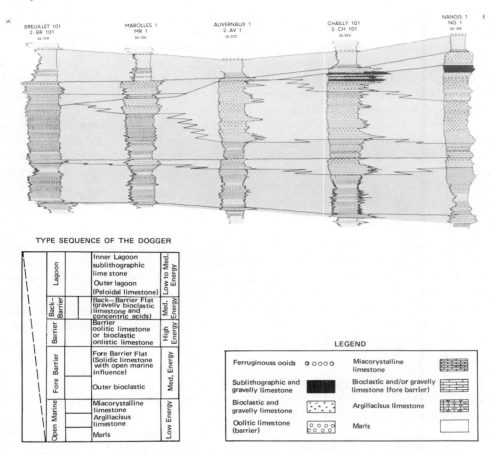

Figure 3-23   Sedimentary sequence interpreted on the basis of well-log analysis.

cement. This kind of cement reduces but does not obliterate both the porosity and the vertical permeability.

Early diagenesis is varied and includes a number of processes that potentially reduce reservoir porosity. Cement may rim particles, either as a fibrous or a granular fringe. The fibrous cement appears to have been similar to the aragonitic or high-Mg calcite cements found in modern marine sediments, particularly in beachrocks (Fig. 2-9c). The granular cement may have a similar origin.

These kinds of early cements can reduce porosity and permeability in parts of the reservoir. They are evenly distributed, but they never entirely obliterate porosity.

Early compaction can reduce void space and produce particle deformation, especially with oöids (Cussey and Friedman, 1977) (Fig. 2-13a). This compaction can locally reduce porosity and permeability. It appears that the granular rim cement can protect the sediment against compaction to some extent.

A drusy mosaic occurs, cementing the base of some of the reservoir zones. It is not well developed and is the only evidence of early freshwater cementation (Fig. 2-9b).

Rim cement is common in all the initially porous facies in the Dogger of the Paris Basin. It developed regularly around echinoderm fragments and its importance depends solely on the abundance and size of these fragments (Fig. 2-14d). A direct link therefore exists between the number and size of echinoderm fragments and the porosity/residual permeability. Rim cement can completely destroy porosity, but does not preclude the occurrence of a nearby undestroyed reservoir.

Coarse spar is an important feature in that it destroys large sections of the reservoirs (Fig. 2-14d). Its development may be linked to solution-precipitation in the presence of freshwater or to compressional tectonic forces.

Pressure solution appears either in the form of stylolitic contacts between particles, which leads to porosity reduction, or as stylolites cutting through the rock. The latter type affects already cemented rock only and so cannot be used as a source for cement in the rocks in which the stylolites occur. The carbonate liberated can be used as a cement some distance from where it is generated.

Conclusions

The paleogeographic reconstruction of the Dogger deposits stresses the importance of interdisciplinary collaboration. This is especially true between the sedimentologists who provide the guidelines for interpreting and understanding the deposits and the log analysts who provide the rapid interpretation by using subsurface data. The results are synthesized, and microstatigraphy is used to check the stratigraphic and ecological correlations. A knowledge of the occurrence of carbonate reservoirs then allows their occurrence and extent to be predicted.

EXAMPLE 5.   DIAGENETIC RESERVOIRS IN REGRESSIVE-TRANSGRESSIVE SEQUENCES, JURASSIC, PROVENCE, SOUTHEASTERN FRANCE

A F. Baudrimont and R. Cussey

Jurassic sequences, composed predominantly of carbonates, occur in the southeastern border area of the Provence in France (Fig. 3-24). The depositional environment of the sequences has been reinterpreted using lithological, well-logging, and sedimentological methods, and correlations have been established between the shelf-edge environment and the basinal deposits.

In this study conventional chronostratigraphic methods were of varying use. In the basin environment, the sequences could be well dated using ammonites. The shelf-edge environment was characterized by facies fossils that are of limited use in dating. This inability to date deposits accurately led to some confusion between the detailed lithostratigraphic units and the uncertain chronostratigraphic subdivisions.

Lithofacies

The sequence from Provence was subdivided into lithostratigraphic units based on well-logging data. The different responses of the logs are based on the varying amounts of carbonate and clay in a deposit. Three major carbonate units have been identified in the basin sequence (Fig. 3-25) belonging to the Lias, Dogger, and Upper Malm. They are separated from each other by argillaceous limestone layers (Liassic Marls and Upper Jurassic "terre noires").

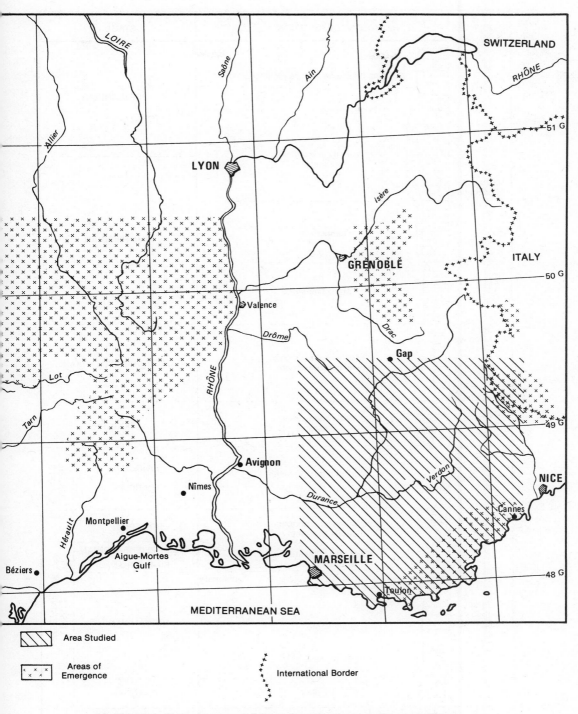

Figure 3-24 Location map of Jurassic sequences, southeastern France.

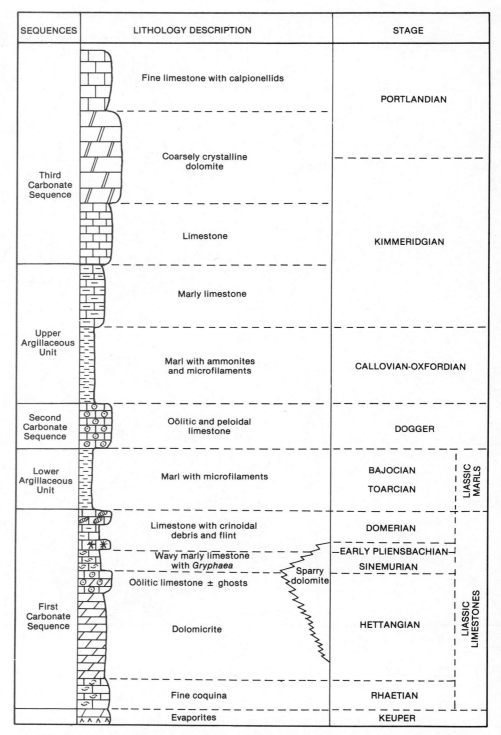

SEQUENCES	LITHOLOGY DESCRIPTION	STAGE	
Third Carbonate Sequence	Fine limestone with calpionellids	PORTLANDIAN	
	Coarsely crystalline dolomite		
	Limestone	KIMMERIDGIAN	
Upper Argillaceous Unit	Marly limestone		
	Marl with ammonites and microfilaments	CALLOVIAN-OXFORDIAN	
Second Carbonate Sequence	Oölitic and peloidal limestone	DOGGER	
Lower Argillaceous Unit	Marl with microfilaments	BAJOCIAN / TOARCIAN	LIASSIC MARLS
First Carbonate Sequence	Limestone with crinoidal debris and flint	DOMERIAN	LIASSIC LIMESTONES
	Wavy marly limestone with *Gryphaea*	EARLY PLIENSBACHIAN / SINEMURIAN	
	Oölitic limestone ± ghosts	Sparry dolomite	
	Dolomicrite	HETTANGIAN	
	Fine coquina	RHAETIAN	
	Evaporites	KEUPER	

Figure 3-25 Basic lithological units of the Jurassic sequence, southeastern France.

132

Example 5   133

Each of the carbonate units, termed the First, Second, and Third Carbonate Sequence respectively, represents part of the overall sequence that evolved in the basin. The main marker horizons shown by the well logs are continuous throughout the study area and probably represent disconformity surfaces, since they always separate two distinct lithologies—the argillaceous limestones and the carbonate units. At the edge of the depositional area the argillaceous limestones are commonly thin or may be completely absent. The carbonate horizons in these areas are also thin or may coalesce to form a sequence of hardgrounds.

From the correlation of these formations and from their distribution in the Provence area the sedimentary evolution of the basin and the position of its margins can be deduced. A study of the lithofacies allows the sedimentary sequences in equivalent deposits to be established without the need for accurate dating.

Using the sequences defined by the stratigraphic study a detailed laboratory analysis was made of the various facies and their depositional environments. No additional field studies were made, so that the interpretations were based on field descriptions and well data, including numerous thin sections. From this study the facies have been defined and the depositional environments interpreted. An attempt has also been made to establish the vertical and horizontal relationships between sediment bodies.

Environmental Interpretations

From information about the distribution of organisms and microorganisms, the faunal associations, the energy of deposition, nonorganic elements present, sedimentary structures, and successions of facies, the environments of deposition have been reconstructed (Fig. 3-26).

Deposits of the supratidal environment are low-energy or completely stagnant mudstones and the environment is characterized by the following features: laminated dolomicrite, evaporite pseudomorphs, fenestral fabrics, shrinkage or solution-collapse breccias, and mud cracks. The fauna is characteristically sparse, and includes thin-shelled gastropods and ostracodes, crustacean coprolites, algal structures, and peloids.

The intertidal deposits are low- to moderate-energy packstones that are characterized by the presence of intraclasts, peloids, dispersed oöids, bioclastic accumulations, benthonic foraminifera, green algae, and stromatolites. These deposits may be dolomitized to form dolomicrosparites and coarsely crystalline dolosparites.

The subtidal deposits inside the barrier consist of low-energy wackestones and mudstones. They contain intraclasts, green algae, recrystallized fine skeletal debris, whole shells, and large benthonic foraminifera. These deposits can also show the effects of dolomitization.

The barrier deposits are high-energy grainstones and boundstones containing abundant oöids, corals, bryozoans, and echinoderms. In places they have been dolomitized to form coarse dolomites, composed of either interlocking crystals or sparse rhombs.

The high-energy shelf grainstones contain bryozoans, corals, echinoderms (crinoids), spicules, peloids, and ferrugenous oöids. These deposits may be locally dolomitized, most commonly with scattered rhombs. The shelf sequence is arranged in accretionary cycles of the type described by Klüpfel (1917).

The low-energy shelf deposits are wackestones and mudstones composed of particles similar to those found in the high-energy deposits, with the addition of

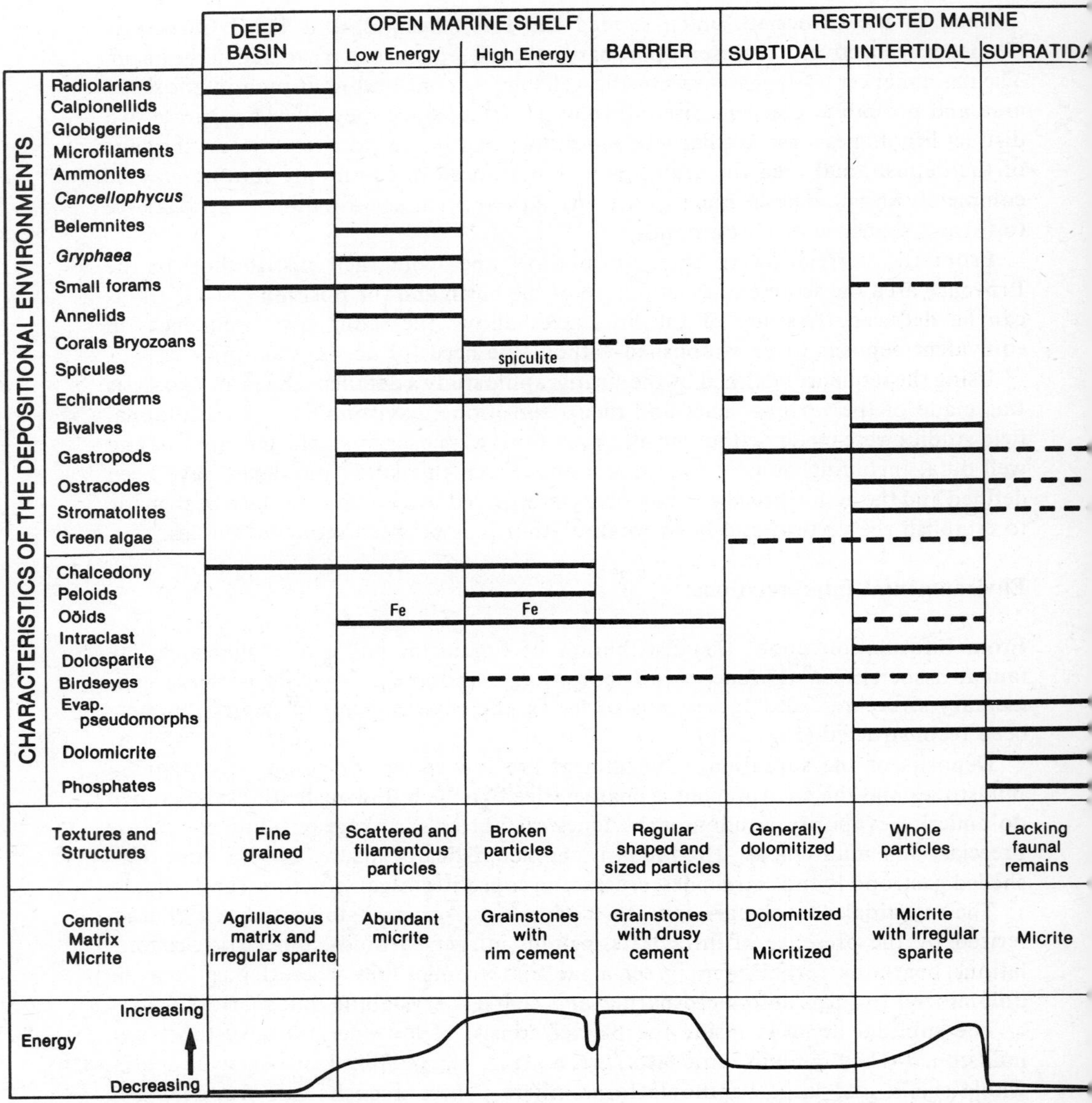

CHARACTERISTICS OF THE DEPOSITIONAL ENVIRONMENTS	DEEP BASIN	OPEN MARINE SHELF Low Energy	High Energy	BARRIER	RESTRICTED MARINE SUBTIDAL	INTERTIDAL	SUPRATIDAL
Radiolarians							
Calpionellids							
Globigerinids							
Microfilaments							
Ammonites							
Cancellophycus							
Belemnites							
Gryphaea							
Small forams							
Annelids							
Corals Bryozoans							
Spicules			spiculite				
Echinoderms							
Bivalves							
Gastropods							
Ostracodes							
Stromatolites							
Green algae							
Chalcedony							
Peloids							
Oöids		Fe	Fe				
Intraclast							
Dolosparite							
Birdseyes							
Evap. pseudomorphs							
Dolomicrite							
Phosphates							
Textures and Structures	Fine grained	Scattered and filamentous particles	Broken particles	Regular shaped and sized particles	Generally dolomitized	Whole particles	Lacking faunal remains
Cement Matrix Micrite	Agrillaceous matrix and irregular sparite	Abundant micrite	Grainstones with rim cement	Grainstones with drusy cement	Dolomitized Micritized	Micrite with irregular sparite	Micrite
Energy Increasing ↑ Decreasing							

Figure 3-26 Characteristics used for interpreting the various depositional environments, Jurassic of southeastern France.

Gryphaea, annelids, belemnites, ammonites, radiolarians, microfilaments, *Cancellophycus,* and calpionellids.

The open marine basin deposits are almost entirely low-energy deposits similar to those of the shelf, but with more common radiolarians, calpionellids, and other pelagic organism. The basinal deposits are arranged as shown in Figure 1-19.

Vertical and Lateral Sequences

Throughout the Jurassic the vertical sequences of deposits that were forming in the open marine environments changed with time (Fig. 3-27). The shelf deposits can be

Example 5 135

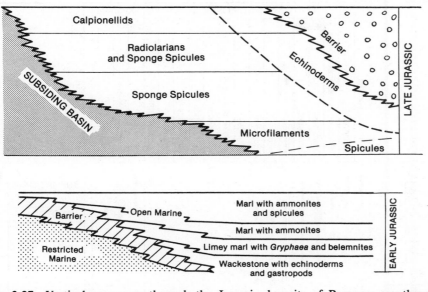

Figure 3-27 Vertical sequences through the Jurassic deposits of Provence, southeastern France.

subdivided into a number of megasequences and cycles. Three phases can be distinguished: Phase 1 is a transgressive megasequence in the Lias (Fig. 3-25), Phase 2 is a regressive-transgressive cycle in the Middle Jurassic (Fig. 3-28), and Phase 3 is an Upper Jurassic regressive megasequence (Fig. 3-28).

These transgressive and regressive phases either do not occur or are difficult to recognize in the basin, since the relative sea-level variations barely affect deep-water deposits. Deposits of the deep basin are only sensitive to climatic changes, especially changes in fauna, and to tectonic changes that may alter sediment transport direction.

The lateral variation in facies is obvious in certain sections and provides additional evidence for the observed megasequences and the lateral displacement of deposits (Figs. 3-29 and 3-30). This large-scale lateral evolution is coupled with small-scale facies oscillations throughout the Jurassic.

At the base of the Lias (Rhaetian) the deposits overlie the Esteral Massif which was probably emergent to the south. Here mudstones, mainly supratidal, and coquinas characterize the tidal flats (Fig. 3-31*a*). Farther north the deposits are more heavily dolomitized and correspond to the outer margin of the shelf. Argillaceous deposits mark the passage to a deeper-water environment, but the lack of field data makes it impossible to establish the presence of basinal deposits. The argillaceous sandstones found associated with these deeper-water deposits have not been studied.

Throughout the Lias most of the shelf sediments were dolomitized (Fig. 3-31*b*). Toward the end of the Lias the deposits became more argillaceous and the shelf was invaded by deeper ocean waters depositing spicule- and echinoderm-bearing sediments. These are the deposits of the first transgressive phase or the so-called First Carbonate Sequence (Fig. 3-31*c*). The shelf-basin boundary was stabilized and remained in the same place until the end of the Jurassic (Figs. 3-31*c* through *g*).

The Dogger deposits indicate higher-energy conditions, and during the regressive phase of the Second Carbonate Sequence restricted marine inner-shelf and barrier deposits were developed (Fig. 3-31*d*). Unlike the Lias, these sediments no longer onlap

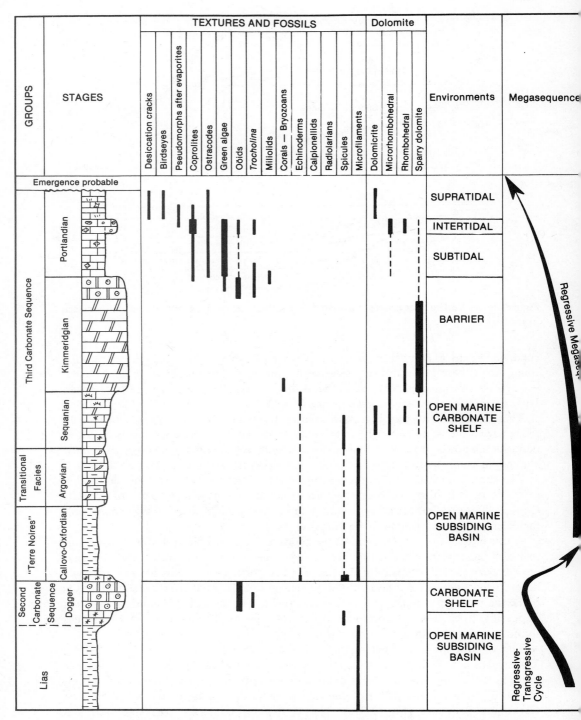

Figure 3-28 Vertical sequences of deposits on the Middle to Upper Jurassic shelf, Provence, southeastern France.

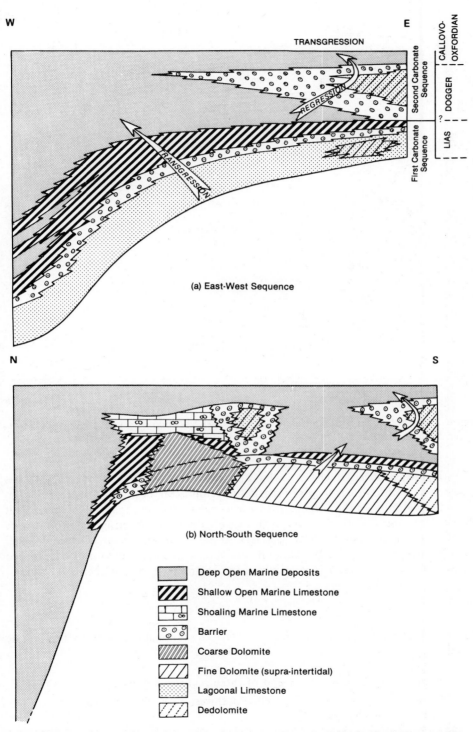

Figure 3-29 Lower and Middle Jurassic facies and vertical sequences, Provence, southeastern France.

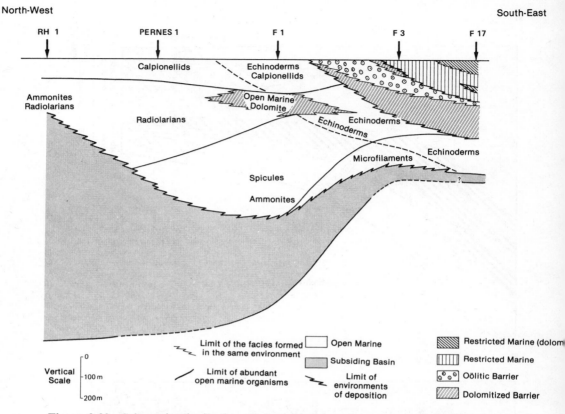

Figure 3-30 Schematic distribution of depositional environments in a SE-NW direction, Upper Jurassic of southeastern France.

the Esterel Massif over which the sea has now passed, but overlie the emergent Maures Massif. The oölitic shoals tend to form deltas reminiscent of those in the Paris Basin.

At the beginning of the Upper Jurassic low-energy deposits again covered the shelf (Fig. 3-31e), and during the Middle Jurassic calpionellid-bearing limestones were deposited throughout the basin (Fig. 3-31f). At the end of the Upper Jurassic the shelf again became an area of high-energy deposition, with the formation of barrier and inner-shelf deposits (Fig. 3-31g), marking the beginning of a new regressive phase.

Conclusions

In southeastern France the following characterized sedimentation during the Jurassic period:

1   Carbonate sequences mark the end of three regressive phases.

2   A gulf-like basin occurred along the southwest-northeast axis and opened out toward the central Alps.

3   Three successive phases of shelf sedimentation are recognized and include a transgression, a regressive-transgressive cycle, and a regression.

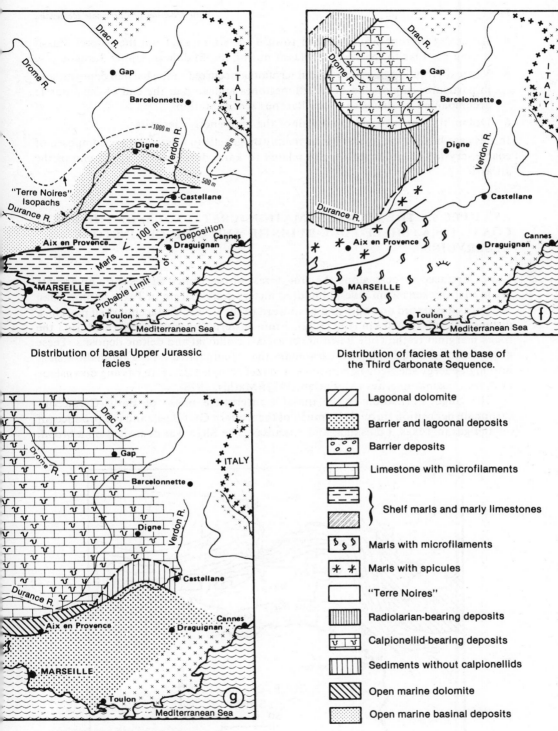

Distribution of basal Upper Jurassic
facies

Distribution of facies at the base of
the Third Carbonate Sequence.

Distribution of facies at the top of
the Third Carbonate Sequence.

Legend:

- Lagoonal dolomite
- Barrier and lagoonal deposits
- Barrier deposits
- Limestone with microfilaments
- Shelf marls and marly limestones
- Marls with microfilaments
- Marls with spicules
- "Terre Noires"
- Radiolarian-bearing deposits
- Calpionellid-bearing deposits
- Sediments without calpionellids
- Open marine dolomite
- Open marine basinal deposits

Figure 3-31 Paleogeographic maps showing distribution of Jurassic depositional environments at margin of the basin, Provence, southeastern France.

139

4 An emergent zone was displaced toward the southeast from the Esterel Massif during the Lias to the Maures Massif during the Middle to Upper Jurassic.

5 Vertical evolution of macro- and microfauna occurred in the basinal deposits, and in particular calpionellid-bearing limestones appeared at the end of the Jurassic accompanied by decrease in argillaceous sedimentation.

6 Dolomitization affected almost all of the high-energy deposits.

In this case history the reservoirs are mainly diagenetic in origin and are composed of coarse, crystalline dolomites or are related to karst development at the end of the Jurassic.

EXAMPLE 6. THE SLIGO FORMATION (CRETACEOUS) OF THE GULF COAST, U.S.A.: SHELF-EDGE RUDISTID AND OÖID-GRAINSTONE RESERVOIRS

Lower Cretaceous rocks along the Gulf Coast of the United States form an arcuate prism that thickens from several hundred meters at its margin in Texas to more than 3000 m as it is traced downdip and basinward to the Lower Cretaceous shelf margin, a distance of 160 to 480 km (Fig. 3-32). From eastern Texas to northern Florida the rocks marginal to the Gulf Basin are siliciclastic alluvial and deltaic deposits. These grade basinward through shelf carbonates that accumulated on a wide shallow shelf into high-energy shelf-edge grainstones and reef complexes before passing downslope into deep basinal micrites (McFarlan, 1977; Mathis, 1978).

The Sligo Formation is the initial transgressive phase of Early Cretaceous carbonate deposition throughout much of the western Gulf Basin and is a producer of oil and gas in Texas, Louisiana, and Arkansas. The Sligo was deposited on a broad,

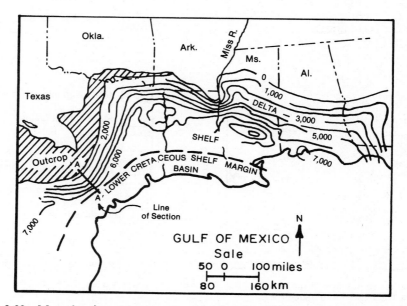

Figure 3-32 Map showing outcrop area and subsurface thickness in feet of Lower Cretaceous strata in the Gulf Coast (after McFarlan, 1977).

Example 6   141

open shelf under conditions that ranged from shallow marine to supratidal. Updip it grades into clastic sandstones and shales, whereas downdip it terminates along the shelf edge and forms a belt of rudistid reefs and oöid grainstone bars (Bebout and Schatzinger, 1978). Hydrocarbon production is from these reefs and grainstone complexes.

Geological Setting

The Sligo Formation is Early Cretaceous (Aptian) in age. It overlies gradationally and is in part laterally equivalent to clastic sandstones and shales of the Hosston Formation (Fig. 3-33). The transitional zone between the Hosston and Sligo represents gradual marine transgression and is marked by a gradation from pure clastics into mixed clastics and limestones and finally into the pure limestones of the upper Sligo Formation.

The Sligo Formation ranges in thickness from less than 15 m in updip sections to over 300 m at the shelf edge. It occurs mostly in the subsurface. Although the Sligo Formation is a transgressive megasequence, relative fluctuations in sea level can be identified on a small scale within the formation, and these are shown by lateral shifts in facies boundaries (Mathis, 1978).

The Sligo Formation is overlain conformably and abruptly by deep open-shelf shales and micrites of the Pine Island Formation, attesting to continued marine transgression during the Early Cretaceous (Bushaw, 1968).

The shelf edge during Sligo times was controlled by the presence of a structural hinge, along which high rudistid reefs and oöid shoals developed.

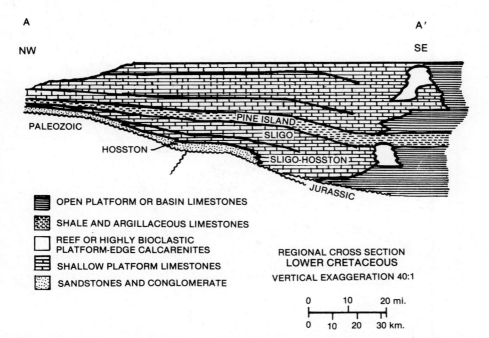

Figure 3-33 Regional cross section through the Lower Cretaceous of the Gulf Coast (after Cook, 1979). Line of section shown in Figure 3-32.

Facies Development and Distribution

The Sligo Formation is a rock unit composed of varying lithologies which was deposited across a shallow shelf (Fig. 3-34). Close to the paleoshoreline the formation consists of gray to brown shales, micrites, limey shales, and dark gray oölitic and fossiliferous limestones and shales that are interpreted as shallow shelf to lagoonal deposits. These pass shoreward into clastic beach and deltaic deposits of the Hosston Formation (Herrmann, 1971).

Sligo lithologies close to the Lower Cretaceous shelf edge include oösparites, oömicrites, coated-grain calcarenites, and broken shell biosparites which interfinger with the argillaceous limestones and shales of the lagoonal sequence. These higher-energy grainstones and packstones occur flanking the rudistid biosparites and biomicrites of the reef trend (Herrmann, 1971). Particles composing the flanking beds include oöids, rudistid fragments, algal pisolites, miliolids, orbitoidids, stromatoporoids, molluscs, echinoderms, algae, bryozoans, ostracodes, and serpulid worms (Mathis, 1978). Some of the rudistid bioherms are up to 60 and even 140 m thick and have a discontinuous distribution along the postulated Early Cretaceous shelf edge as well as on the inner shelf in central Louisiana (Fig. 3-34).

Shelf-edge and inner-shelf high-energy buildups include oöid bars and banks that have an anastomosing trend (Herrmann, 1971). The shelf-edge oöid grainstone bars may be flanked on the inner-shelf side by grain-coated packstones, peloidal packstones and grainstones, and peloidal intraclast packstones. On the outer shelf the oöids are

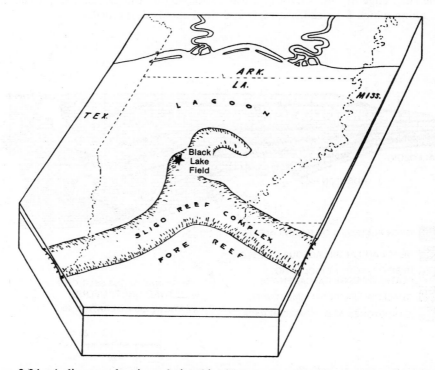

Figure 3-34 A diagram showing relationships between postulated environments and the position of the reef trend during Sligo times (after Herrmann, 1971).

Example 6   **143**

flanked by skeletal grainstones that grade basinward into outer-shelf wackestones and mudstones (Bebout and Schatzinger, 1978).

Dolomite and minor evaporites occur within the Sligo Formation, but their distribution is apparently complex and without obvious trend. Dolomite may form up to 60% of the formation on the southern side of the reef trend in Louisiana, whereas it is most common in the back-reef or back-bar deposits in Texas (Bushaw, 1968; Herrmann, 1971).

The Black Lake Field

Before the discovery of the Black Lake field in the Sligo rudistid reef limestone in central Louisiana (Fig. 3-34), most of the oil and gas production from the Sligo was out of anastomosing oölitic bars and banks.

The Black Creek field is one of the largest Lower Cretaceous carbonate hydro-carbon reservoirs on the U.S. Gulf Coast and it produces from a porous zone composed of a variety of carbonate lithologies in the upper part of the Sligo Formation (Fig. 3-35 and Table 3-3).

In a study of the Black Creek field by Mathis (1978), eight major carbonate lithofacies were recognized: (1) rudistid (caprinid), (2) oncolitic, (3) oölitic, (4) bioclastic, (5) bioclastic micritic, (6) foraminiferal, (7) micritic, and (8) dolostone. The distribution of these lithofacies in a section across the field taken perpendicular to depositional strike is shown in Figure 3-35.

The rudistid lithofacies is composed of micritic rudistid and micritic rudistid skeletal wackestones, packstones, and minor boundstones in which the rudistids may be upright, encumbent, whole, or broken. In the wells studied this lithofacies had a maximum thickness of 18 m and a linear trend across the field. Packstone textures are more common on the seaward side of the trend, whereas on the landward side wackestones predominate.

Bioclastic and bioclastic micritic limestones are adjacent to and overlie the rudistid lithofacies. Bioclastic micritic and micritic limestones underlie the rudistid biofacies and pass gradationally upward into it.

The rudistid reef in the Black Lake field is believed to have formed a wave-resistant barrier which developed together with the laterally adjacent facies during a period of gradual marine transgression.

Distribution of porosity and permeability in the lithofacies of the Black Lake field are shown in Figure 3-36. Porosity in the rudistid lithofacies ranges from 2 to 25%, with an average of 13%. The permeability ranges from 0 to 1040 md with an average of 65 md. Pore types present include moldic, vug, fracture, interparticle, and intraparticle, which may have been modified by solution or cementation.

Several of the associated lithofacies have significant porosity and permeability. The oncolite lithofacies, which may be up to 8.8 m thick in the section studied, has porosities that range from 2 to 28% (average 8%) and permeabilities of 0 to 294 md (average 28 md). The oölite lithofacies is up to 10 m thick and has porosities that range from 5 to 27% and permeabilities of 0 to 2285 md, with an average of 211 md. The bioclastic lithofacies has porosities of 2 to 28% (average 14%) and permeabilities of 0 to 1975 md, with an average value of 74 md. Finally, the dolostone lithofacies, which is only a maximum of 2 m thick, has porosities of 5 to 17% and permeabilities of 0 to 52 md, with an average of 11 md (Mathis, 1978).

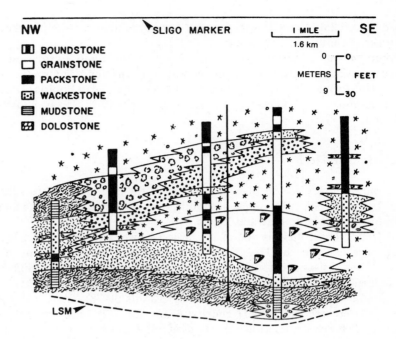

NW   SLIGO MARKER   I MILE   SE

1.6 km

- BOUNDSTONE
- GRAINSTONE
- PACKSTONE
- WACKESTONE
- MUDSTONE
- DOLOSTONE

METERS   FEET

0 — 0

9 — 30

LSM

LITHOFACIES KEY

- RUDISTID (CAPRINID)
- ONCOLITE
- OÖLITE
- BIOCLASTIC
- BIOCLASTIC MICRITIC
- FORAMINIFERAL
- MICRITE
- DOLOSTONE

Figure 3-35 Stratigraphic cross section perpendicular to depositional strike across the Black Lake field showing the distribution of the eight lithofacies and their textures (from Mathis, 1978).

144

Example 6 145

Table 3-3 Basic Reservoir Data for the Black Lake Field, Central Louisiana (from Mathis, 1978)

General Reservoir Data

Producing Formation	Sligo (Pettet)
Average Producing Depth (feet)	7,950
Productive Area (acres)	16,940
Original Gas-Oil Contact (feet below sea level)	7,835
Original Oil-Water Contact (feet below sea level)	7,870
Original Reservoir Pressure (psig)	4,020
Reservoir Temperature (°F)	243
Average Gas Zone Thickness (productive feet)	51
Average Oil Zone Thickness (productive feet)	22

Reservoir Rock Properties

	Gas Zone	Oil Zone
Average Porosity (percent)	16.3	15.6
Average Permeability (millidarcys)	133.0	74.0
Average Water Saturation (percent)	30.1	37.4

Reservoir Fluid Properties

Oil Gravity (°API)	45.8
Specific Gravity Gas	0.716
Formation Water Salinity (ppm NaCl)	150,000

Reservoir Volumes

Net Oil Zone Rock Volume (acre-feet)	220,147
Net Gas Zone Rock Volume (acre-feet)	737,379
Original Oil in Place (stock-tank barrels)	105,360,000
Original Gas in Place (million standard cubic feet)	788,775
Original Condensate Content of Gas Zone (stock-tank barrels)	50,797,000

Source: Core Laboratories, Inc., 1965

Diagenetic alterations, both synsedimentary and postdepositional, have altered the original porosity and permeability values for the various lithofacies. Alterations detrimental to porosity included compaction, pressure solution, internal sedimentation, precipitation of synsedimentary bladed rim cement, late diagenetic void filling by syntaxial overgrowths of calcite, equant calcite cement, saddle dolomite, anhydrite, and kaolinite. Permeability has probably been enhanced to some extent by both pre- and postlithification fracturing and solution.

Conclusions

The Sligo Formation is a transgressive megasequence representing the initiation of carbonate deposition during Early Cretaceous time in the Gulf Coast region. The formation is composed of a variety of lithologies deposited over a broad shelf and passing shoreward into clastic deltaic and alluvial deposits.

From the paleoshoreline out into the depositional basin the Sligo Formation consists of mixed clastics and carbonates which grade into micrites, limey shales, and dark gray oölitic and fossiliferous limestones interpreted as shallow shelf to lagoonal

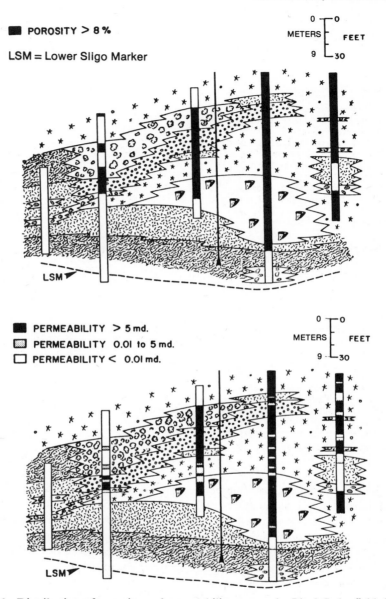

Figure 3-36 Distribution of porosity and permeability across the Black Lake field. For key to lithofacies and horizontal scale see Figure 3-35 (from Mathis, 1978).

deposits. On the shelf edge these lithologies interfinger with oölitic limestones and calcarenites which are marginal to the shelf-edge rudistid buildups. These reefs are noncontinuous and grade laterally into shelf-edge oöid banks and bars.

Before the discovery of the Black Lake field hydrocarbon production from the Sligo was restricted to the inner-shelf and shelf-edge oöid banks and bars. The Black Lake field occurs to the north and shoreward of the major reef trend and produces from a bioherm of porous rudistid limestones and associated oölitic, bioclastic, and oncolitic deposits.

Example 7   147

EXAMPLE 7.   REEF RESERVOIRS FROM THE MIDDLE DEVONIAN OF NORTHERN ALBERTA, CANADA

E. Elloy

A reef does not necessarily constitute a reservoir because reefs are not always porous. Reefs form reservoir horizons that are not necessarily better than any other lithology that is found in a similar paleogeographic setting or has undergone favorable diagenetic alteration. When diagenetic evolution and paleogeographic position are favorable for reservoir formation the following aspects may make reefs important potential reservoirs:

1   Some reef rock has very large pore spaces.
2   If no argillaceous layers obstruct drainage of fluids, particularly in a vertical direction, reef rock tends to become homogeneous in its properties.

These two points mean that a reef reservoir may have high productivity and only a few wells are needed to drain the entire reef body. Oil saturation is significant, in the order of 85 to 90%, and the possibility of high primary recovery of *in situ* oil—around 40 to 50%—makes such a reservoir an attractive exploration target.

Reefs are stratigraphic traps that are more confined in time and space than a majority of sedimentary rock bodies. Their location in the marine environment attracts abundant flora and fauna, and they are surrounded and enclosed by diverse and different kinds of sedimentary rocks. Reefs also occur in close association with beds rich in organic matter. These two points are important, since the concept of a reservoir includes both the presence of a source rock for the hydrocarbons and a caprock or seal that will prevent their escape.

Rainbow Reefs, Alberta, Canada

Reefs from the Rainbow Field, Alberta, Canada, developed during the Middle Devonian Givetian Stage in a small intracratonic basin or subbasin separated from the Rocky Mountain Geosyncline by a carbonate barrier (Fig. 3-37). These reefal bodies of various geometries include atolls with associated lagoonal facies, arcuate atolls, patch reefs, and smaller pinnacle reefs (Fig. 3-38). About 40 reef buildups of various sizes are known, the largest about 6 km long and the smallest measurable in meters. The largest buildup projected close to 300 m above the seafloor.

Four different combinations of fluids are found in the Rainbow reef buildups: (1) gas and water, (2) gas and oil, (3) oil and water, and (4) water.

Both reefs A and B (Fig. 3-38), which are atolls, and reef F, the relatively nonporous but largest buildup in the field, contain the volume of stored fluids that has been computed as follows:

> Reef A   Area = 400 hectares, oil column = 130 m, gas column = 75 m
> Oil volume = 25×10^6 m^3
> Gas volume = 740×10^6 m^3
>
> Reef B   Area = 1700 hectares, oil column = 108 m
> Oil volume = 45×10^6 m^3

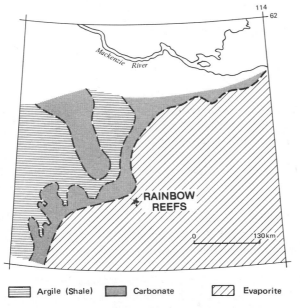

Figure 3-37 Location map and paleogeographic position of the Rainbow reefs in the Middle Devonian of western Canada.

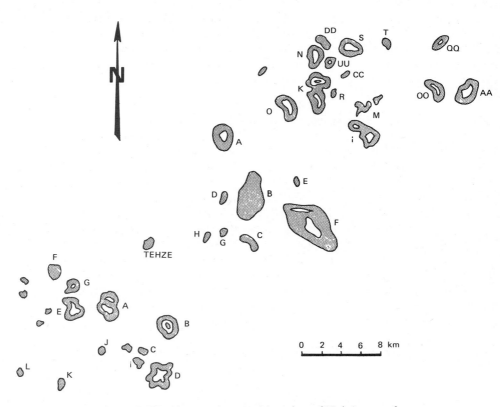

Figure 3-38 Shapes, sizes, and location of Rainbow reefs.

Example 7 149

Reef F Area = 1950 hectares, oil column = 90 m, gas column = 40 m
Oil volume = 33×10^6 m^3
Gas volume = 1700×10^6 m^3

Lithostratigraphy

The granitic basement is overlain by Early Devonian to Early Middle Devonian granitic sand followed by strata of evaporites (anhydrite). Figure 3-39 shows the overlying carbonate unit which is of constant thickness and lithology. This unit, the Lower Keg River Formation, acts as a base to the reef buildups. Reef growth, the Rainbow Member, and the associated organic-rich sediments of the Upper Keg River Formation are covered by a second sequence of evaporitic strata. Increasing restriction within the basin has resulted in deposition of salt (Black Creek Member) followed by anhydrite and dolomite of the Muskeg Formation. This vertical sequence represents one of the best known examples of closed-basin evaporites (Fig. 3-39).

Faunal and lithological zonation is seen in the reefs, and the rocks that occur are best classified according to Embry and Klovan (1971) in addition to Dunham (1962) (Table 3-4). From the basal carbonate sheet underlying the reef buildup the vertical sequence developed is as follows:

1 Dark micrites and bafflestones with a particular assemblage of bryozoans, algae, "stromatactis," and fasciculate tabulate corals.

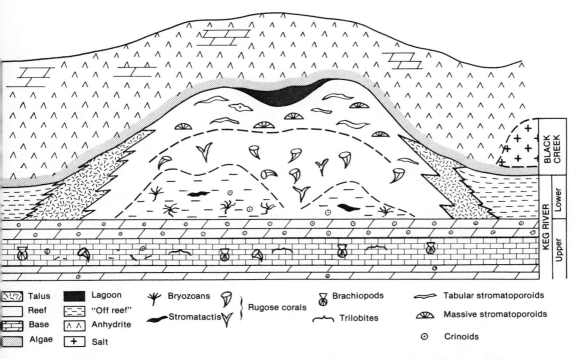

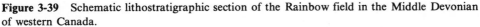

Figure 3-39 Schematic lithostratigraphic section of the Rainbow field in the Middle Devonian of western Canada.

Table 3-4 Classification of Embry and Klovan (1971) Complementing Dunham's (1962) Classification.

Allochthonous Limestones Original components not bound together during deposition						Autochthonous Limestones Original components bound together during deposition		
Less than 10% of particles >2 mm				More than 10% of particles >2 mm		Organisms building a rigid framework	Organisms encrusting and cementing the frame elements	Organisms creating baffles
Contains Lime Mud (<0.03 mm)			Lacks Lime Mud	Allochems >2 mm within matrix	Particle-supported allochems >2 mm			
Lime-Mud Supported		Particle-Supported						
Less than 10% particles >0.03 mm <2 mm	More than 10% particles >0.03 mm <2 mm						Boundstone	
Mudstone	Wackestone	Packstone	Grainstone	Floatstone	Rudstone	Framestone	Bindstone	Bafflestone

Lime-Mud Supported ← → Particle Supported

Less than 10% particles ← → Greater than 10% particles

Allochems

Lime Mud — Lime-Mud Free

Matrix or Cement

Bioaggregates

Dunham	Embry & Klovan
WACKESTONE	→ WACKESTONE → FLOATSTONE
PACKSTONE	→ PACKSTONE → RUDSTONE
GRAINSTONE	→ GRAINSTONE → RUDSTONE

150

Example 7 151

2 Bindstone-framestones (Dunham's boundstones) with rugose and tabulate corals dominated by branching forms.

3 Framestones with massive tabulate corals and stromatoporoids, the latter being more abundant toward the top of the reef.

4 Lagoonal deposits that range from mudstones to boundstones, containing stromatoporoids of the *Stachyodes* and *Amphipora* type.

5 At the top of the buildup algal bindstones have developed.

6 Laterally the reef is surrounded by a talus slope made up of debris of different lithologies (rudstones, floatstones, and wackestones).

A fauna of branching corals developed locally on the talus slope where crinoids are extremely abundant. Farther out from the toe of the reef the sediment facies grades into thin deposits of oozes and micrites rich in organic matter.

This zonation indicates the great variability of facies present in a reef association and thus the different behavior of the rocks toward the confining fluids. The original depositional features of the different facies that dictate this behavior may be modified by recrystallization, dolomitization, and fracturing. This modification is responsible for the heterogeneity found within a single reef.

The beds at the base of the reef are bafflestones composed of branching corals in a micritic matrix. These beds have a lithology similar to so-called mudmounds and, because of a lack of initial porosity, they have unfavorable petrophysical properties.

The crinoidal grainstone facies at the base of the reef would have been an excellent reservoir when deposited, but its porosity has been destroyed by cementation. The cement grew outward from each crinoid fragment as if particle and cement were a single crystal. This epitaxial overgrowth or rim cement occluded the porosity of the crinoidal grainstone. Any superimposed dolomitization has not increased the porosity of the rock. In the most favorable examples this grainstone may have a porosity of 4 to 5% and a permeability between 2 and 5 md.

The talus deposits surrounding the reef have variable but generally good reservoir properties. The talus deposits that occur on the open marine side of the reefs have porosities of 12 to 14%, whereas the talus on the lagoonal side of the reef has porosities of up to 16%.

Except for the lagoonal talus deposits, the deposits of the lagoon are devoid of primary porosity and permeability.

The reservoir characteristics of the reef facies are good to excellent. In Reef A such rocks have a porosity of 12 to 13% and permeabilities of 100 to 200 md. The range of these values can be large. In the same single reef that has an average of 12% porosity and a permeability of 597 md extremes may yield 29.4% porosity and permeabilities up to 1680 md (Table 3-5). The average is established from the total pay zone of the reef.

Table 3-5 Porosity and Permeability of Rainbow Reefs

Reef	Porosity %	Permeability md	
		horizontal	vertical
E	13.2	184	30
A	11.8	597	116
B	7.2	306	42
F	4.7	223	40
K	3.0	—	—

Cores taken every 30 cm in the pay zone of reefs show major variations in permeability values both in a horizontal and vertical direction.

Diagenetic Alteration

The original properties of the rocks that make up the reef associations have been greatly modified by diagenesis; the exact extent of this diagenesis is impossible to estimate. Dolomitization was one of the most common processes that changed the reef rocks drastically. In the Rainbow Field the following kinds of reefs have been recognized: (1) limestone reefs, (2) partially dolomitized reefs, (3) dolomitic reefs, (4) calcareous reefs with partially dolomitized lagoonal deposits, and (5) dolomitic reefs whose lagoonal deposits have escaped dolomitization.

In partially dolomitized reefs the dolomite is always found in the upper part of the reef. This would indicate dolomitization from the surface down by descending fluids.

Dolomitization appears to have improved the reservoir properties of the rock and is thus consistent with the general theory that recognizes increasing pore volume with epigenesis. The most common kinds of pores include:

1 Intercrystalline pores preserved by the smooth faces of the dolomite crystals. The size of these pores varies from 400 to 2000 microns and they have an angular shape.

2 Vugs resulting from solution of calcitic tests or skeletal fragments that have escaped dolomitization. The dolomite crystals may project into the vugs, slightly reducing pore volume. The size and shape of these vugs vary greatly depending on the size and shape of the test or skeletal fragment dissolved.

3 Karstic porosity resulting from extensive solution has removed parts of the reef. Filling by calcite of these large voids occurs in some parts of the rock, reducing porosity.

In many places the tight micrite of the lagoonal facies has developed a secondary porosity as a result of dissolution of *Amphipora,* dolomitization of the micrite, or solution enlargement of bird's-eyes. Similar kinds of porosity have developed in the overlying algal laminites.

In the carbonate sand sheet basal to the reefs the original porosity has been destroyed by epitaxial overgrowths around the crinoid ossicles, and porosity has not been improved by epigenesis. In the overlying deposits (the core of Reef A) there has been slight reduction in porosity coincident with replacement by dolomite. The decrease in porosity shown by dolomitic beds in Reef H and Tehze Reef is shown in Figures 3-40 and 3-41. In Reef H these dolomites formed at the expense of limestones which separate the reef from the lagoonal sediments and the lagoonal sediments from the algal laminites. In Tehze Reef the limestone separating the reef and lagoonal sediments has been dolomitized.

In addition to calcite and dolomite, which can obliterate porosity and block fluid circulation, the following void-filling minerals occur:

1 Sulfates (anhydrite) precipitated mainly at the edge of the reef and were introduced from the surrounding Muskeg Formation.

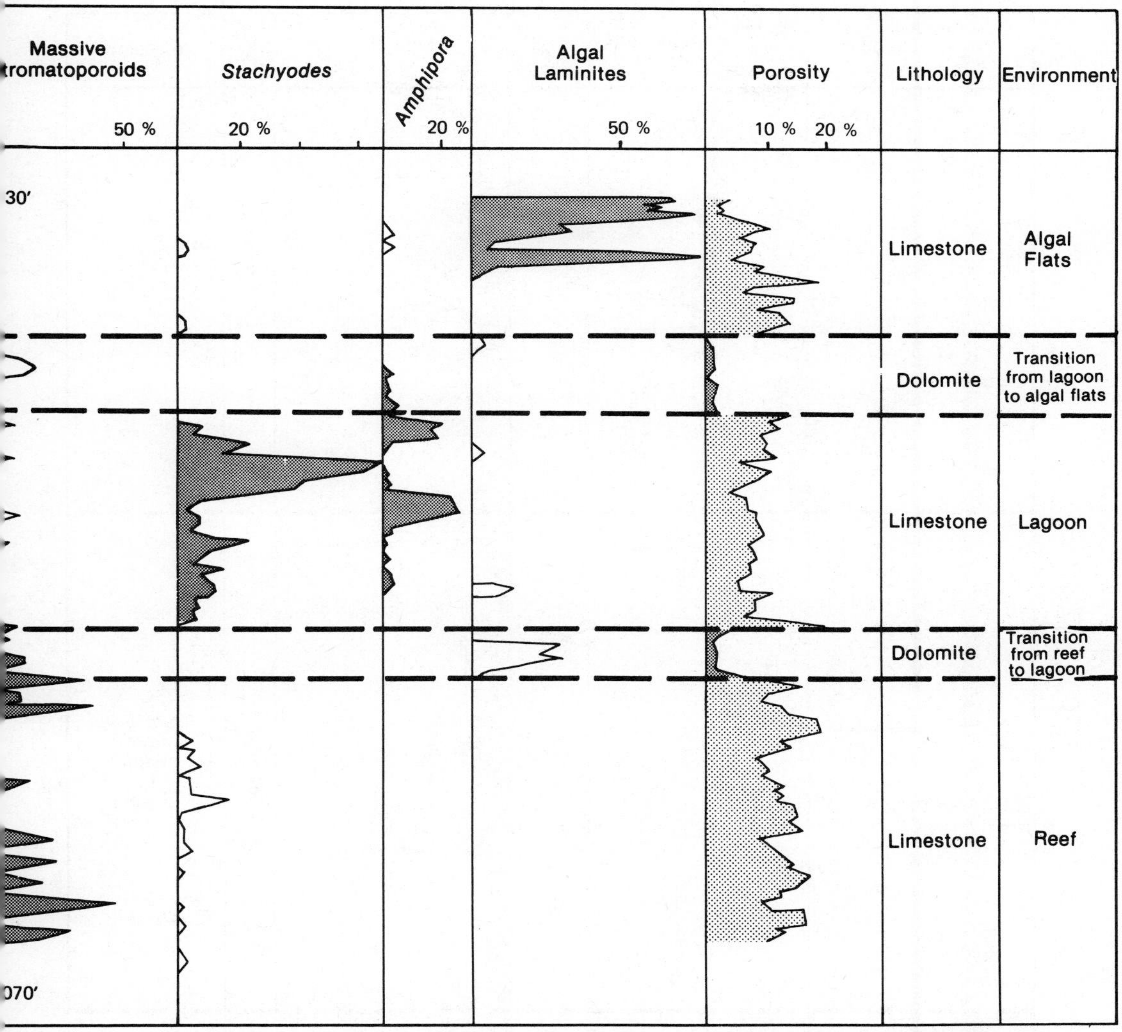

Example 7   153

Figure 3-40 Vertical sequence in Reef H showing relationship between distribution of porosity, flora, fauna, and environment.

2   Fluorite occurs as a pore-filling mineral in the reefal deposits.

3   Sulfides such as sphalerite are distributed in seams several centimeters long.

4   Pyrobitumens occur in parts of the reservoir.

Fracturing is an important consideration in the study of permeability, especially where fractures and fissures form an interconnective network. Small core samples do not provide sufficient material for predicting the behavior of fissures and fractures in the surrounding rock. The distribution and characteristics of fractures, whether the frequency is high or low and whether the fractures are open or healed, are relative only and vary greatly over small distances. Because of this unpredictable behavior fissures

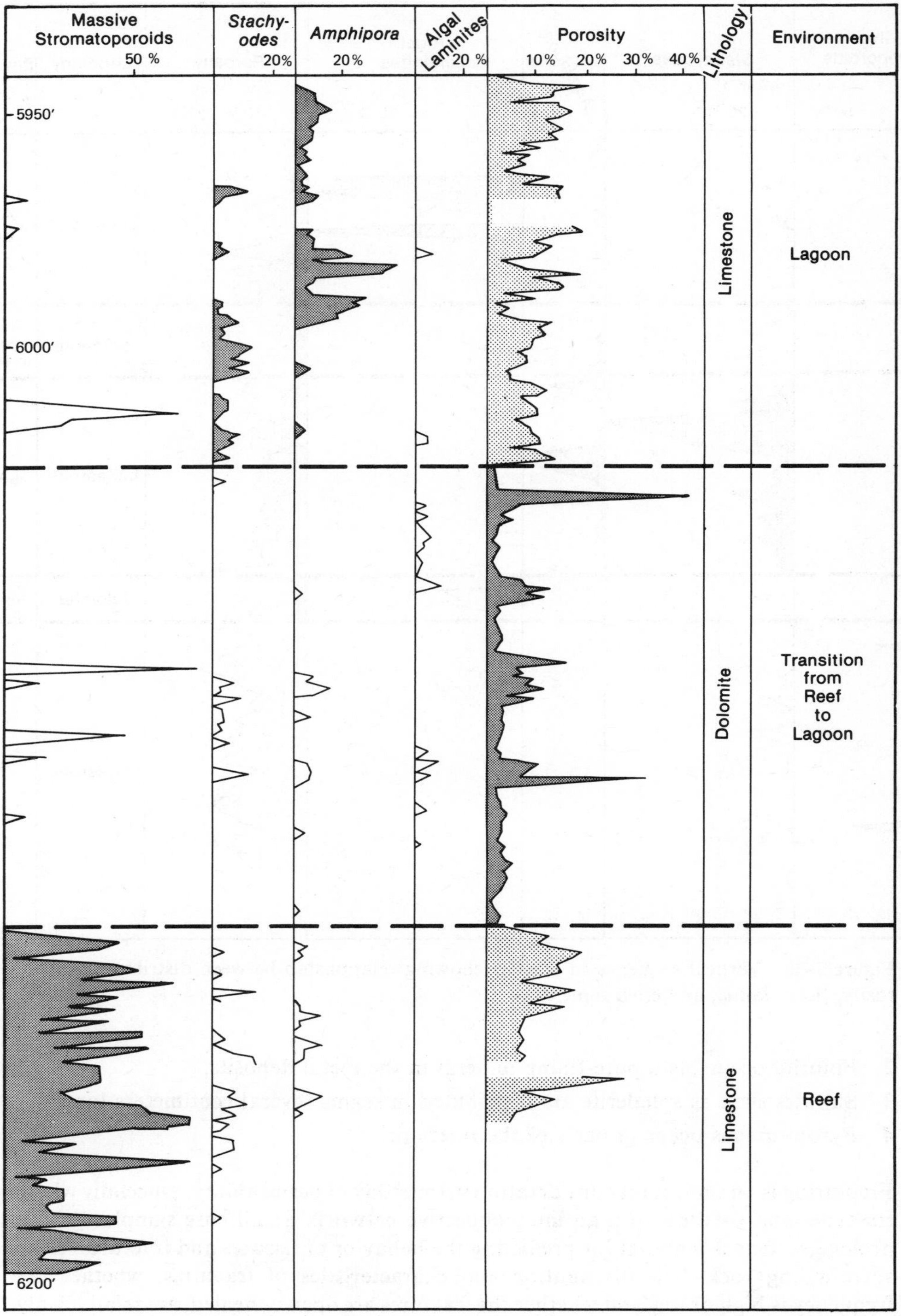

Figure 3-41 Vertical sequence in the Tehze Reef showing relationship between distribution of porosity, lithology, flora, fauna, and environment.

154

Example 7 155

within the reef cannot be correlated with permeability curves. Other important factors to consider include microporosity and other pore interconnections.

Reef F, which is highly fractured, is an exception in the Rainbow Field. Most of the fractured rock occurs in wells on the outer edge of the reef fringe. Wells drilled on the lagoonal side of the reef show less fracturing. The degree of fracturing intensifies toward the surface of the reef, the rock here being more rigid and more susceptible to breakage in response to stress.

Figures 3-42 and 3-43 show the distribution of principal organisms, porosity, and permeability based on seven wells in Reef F. They confirm the erratic distribution of organisms and physical properties in the reef sequence. However, some correlation exists between parameters in the lagoon and the buildup fringe of the reef proper. Figure 3-44 illustrates the extent of fracturing in Reef H, which is greater on the fringes of the reef where it forms the most promising reservoir zones.

Conclusions

The study of the Rainbow reefs illustrates the many disciplines involved in a study of a carbonate-rock reservoir, for example, paleogeography, petrology, petrophysics, and petrography. This study also illustrates the importance of reconstructing the diagenetic history of the sequence, including the effects of fracturing on the petrophysical properties of the rocks.

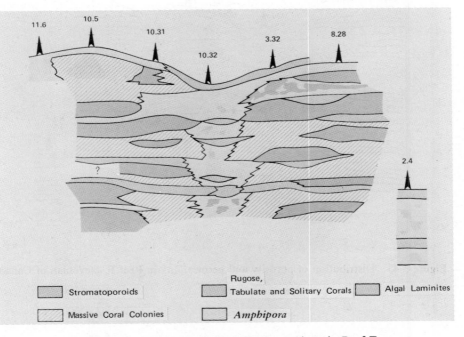

Figure 3-42 Distribution of principal organisms in Reef F.

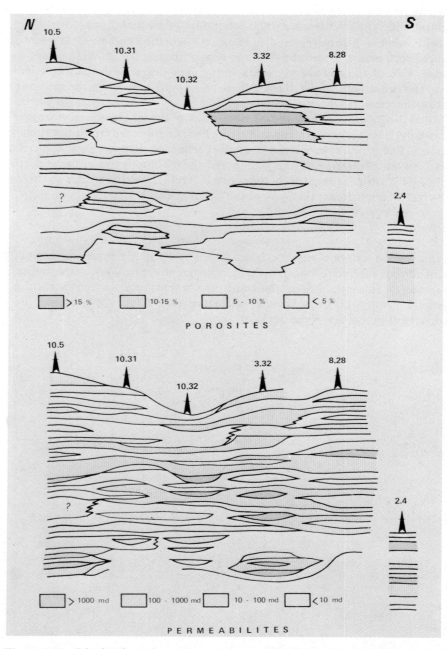

Figure 3-43 Distribution of porosity and permeability in Reef F, Devonian of Canada.

Example 8 157

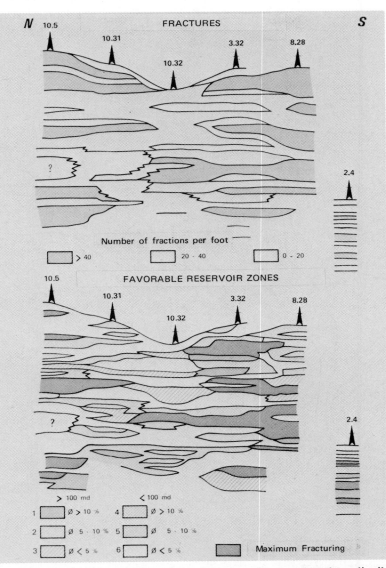

Figure 3-44 Distribution of fractures and favorable reservoir zones based on distribution of permeability, porosity, and fractures, Reef F, Rainbow Field, Devonian of Canada.

EXAMPLE 8. PINNACLE-REEF RESERVOIRS FROM THE MIDDLE SILURIAN OF THE MICHIGAN BASIN, U.S.A.

The Michigan Basin is a shallow, circular, intracratonic sag covering 122,000 square miles in the northern central United States and southern Canada (Fig. 3-45). During Middle and Late Silurian times the basin was the site of extensive carbonate deposition and of precipitation of basin-filling evaporites. The carbonate deposits include basin-margin barrier reef or bank complexes, pinnacle reefs that formed on the shelf offshore from the barrier reefs, and micrites deposited in the center of the basin (Fig. 3-46). The

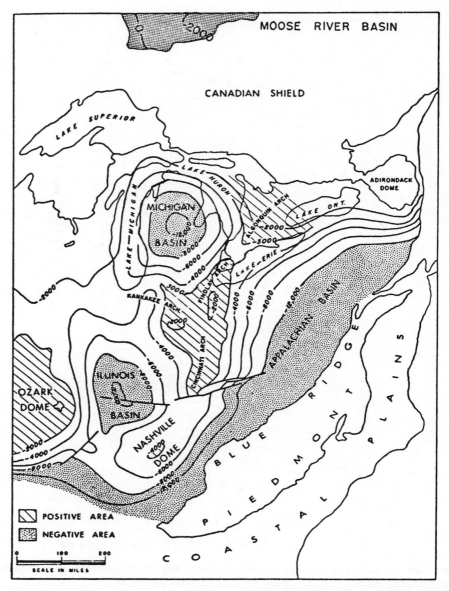

Figure 3-45 Structure contour map of regional basement showing the location of the Michigan Basin. Contour lines represent feet below sea level (from Nurmi, 1975, modified after Brigham, 1971).

major development of carbonates is overlain by extensive, cyclical evaporite/carbonate deposits of Late Silurian age.

Most of the oil and gas reservoirs in the Michigan Basin are found within individual pinnacle reefs. These reefs may be partially dolomitized and commonly have good porosity and permeability as a result of periods of subaerial exposure and freshwater leaching during the Silurian. The pinnacles are sealed both on their flanks and on their tops by the evaporites, and in some reefs evaporites completely plug the intrareef porosity.

Example 8 159

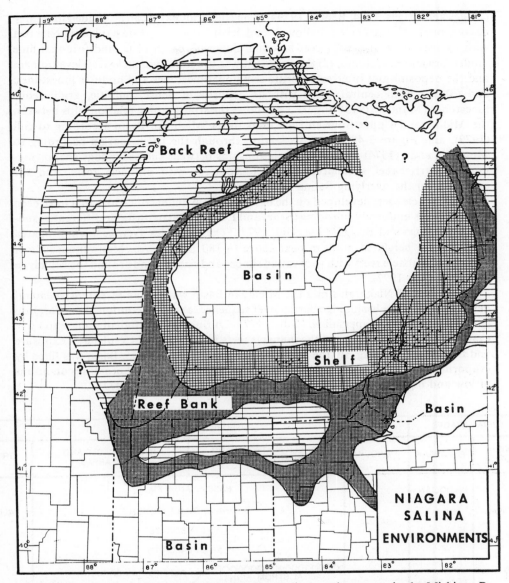

Figure 3-46 Middle and Late Silurian (Niagara-Salina) environments in the Michigan Basin. The pinnacle reefs are located along the shelf zone (from Briggs and Briggs, 1974, modified after Ulteig, 1964; Sanford, 1969; Brigham, 1971; Shaver et al., 1971; Mantek, 1973; Meloy, 1974).

Geological Setting

The Michigan Basin began subsiding in the Precambrian and contains up to 14,000 feet of undeformed Paleozoic strata. Maximum subsidence occurred during the Late Silurian and Middle Devonian (Cohee and Landes, 1958).

During the Niagaran (Middle Silurian) the Michigan Basin was situated near the equator and filled by warm shallow seas that provided ideal conditions for reef growth

(Briggs and Briggs, 1974; Bently, 1979a). The Niagaran Basin can be divided into three depositional settings: (1) the shallow, broad, basin-edge carbonate bank or barrier reef with its reef limestone, back-reef lagoonal deposits, patch reefs, and fore-reef lime mudstones and sandstones; (2) the shelf that was the site of pinnacle-reef development and the deposition of interreef micritic crinoidal limestones and nodular limestones; and (3) the deep central basin with its deposits of dense, micritic, argillaceous limestones (Mantek, 1973). Figure 3-47 shows the stratigraphic relationships between the Middle and Upper Silurian rock units in the Michigan Basin according to Gill (1979), and Figure 3-48 is a cross-section through the basin margin according to Mesolella *et al.* (1974) showing the relationships between the basin-edge shallowing-upward barrier-reef sequence, the shelf sequence with pinnacle reefs, and the evaporites of the overlying Salina Group.

The pinnacle reefs developed on the basin platform under conditions that have been interpreted as uniformly subsiding (Sears and Lucia, 1979) or unstable differential subsidence (Briggs and Briggs, 1974; Gill, 1977). The pinnacles average 0.5 km^2 in area and range in height from 90 m near shore to 180 m offshore. Many thousands of individual pinnacle reefs with very varied shapes and sizes extend around the Michigan Basin (Shaver, 1977).

At the end of Niagaran times the Michigan Basin was barred by extensive lateral development of the basin-edge barrier-reef sequence, and with continued evaporation conditions became hypersaline (Gill, 1977). During this period of time extensive evaporites were deposited and the pinnacle reefs were subjected to subaerial exposure and freshwater leaching. By the end of Salina times the Michigan Basin was filled by evaporites and intercalated thin limestone units which represent alternating conditions of low and high sea level (Nurmi, 1975; Nurmi and Friedman, 1977).

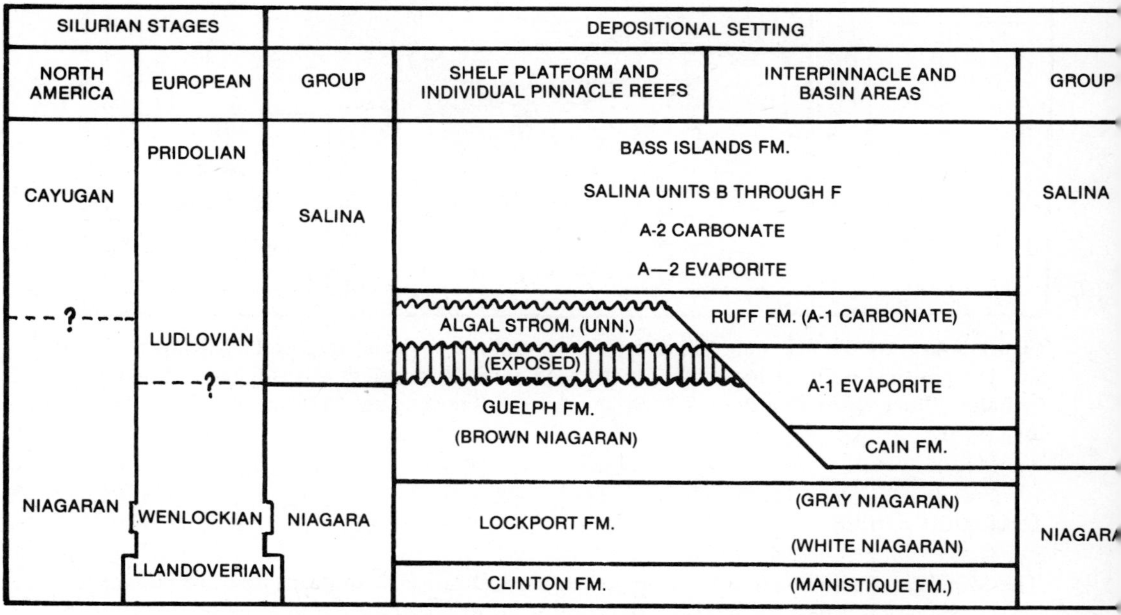

SILURIAN STAGES			DEPOSITIONAL SETTING		
NORTH AMERICA	EUROPEAN	GROUP	SHELF PLATFORM AND INDIVIDUAL PINNACLE REEFS	INTERPINNACLE AND BASIN AREAS	GROUP
CAYUGAN	PRIDOLIAN	SALINA	BASS ISLANDS FM.		SALINA
			SALINA UNITS B THROUGH F		
			A-2 CARBONATE		
			A—2 EVAPORITE		
- - ? - - -	LUDLOVIAN		ALGAL STROM. (UNN.) (EXPOSED)	RUFF FM. (A-1 CARBONATE)	
	- - -?- -			A-1 EVAPORITE	
			GUELPH FM. (BROWN NIAGARAN)		
				CAIN FM.	
NIAGARAN	WENLOCKIAN	NIAGARA	LOCKPORT FM.	(GRAY NIAGARAN) (WHITE NIAGARAN)	NIAGARA
	LLANDOVERIAN		CLINTON FM.	(MANISTIQUE FM.)	

Figure 3-47 Stratigraphic relationships and nomenclature of Middle and Upper Silurian formations in the Michigan Basin (from Gill, 1979).

Example 8   161

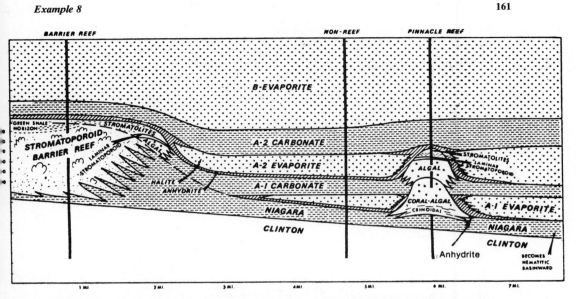

Figure 3-48 Schematic cross section of Middle and Upper Silurian rocks in the Michigan Basin showing the relationship between the basin-margin reefs and shelf pinnacle reefs (after Mesolella et al., 1974).

Much controversy exists in the interpretation of the relationship between pinnacle-reef development and the precipitation of evaporites. Three schools of thought, summarized by Mesolella *et al.* (1974), relate relative timing of reef development and evaporite precipitation:

Model 1   The pinnacle reefs developed in their entirety during Niagaran times and are postdated by the deposition of Salina carbonates and evaporites (see Fig. 3-47)

Model 2   The development of the pinnacles occurred at the same time as the precipitation of the surrounding Salina evaporites

Model 3   Periods of reef development were followed by periods of hypersalinity and precipitation of evaporites. In this model the reefs and evaporites follow each other closely in time through several cycles of deposition, but are not synchronous (see Fig. 3-48)

The Pinnacle Reefs

The development of an individual pinnacle reef has been divided into several stages, but the interpretation of these stages differs depending on the reef/evaporite depositional model that is upheld. In general the pinnacle reefs overlie crinoidal, bryozoan wackestones of the Upper Lockport Formation. The actual pinnacle development has been divided into four stages (Briggs and Briggs, 1974; Huh et al., 1977; Bentley, 1979 a, b).

Stage 1 was the initial development on the shelf of moundlike bodies of carbonate mud mixed with the skeletons of bryozoans and crinoids (Fig. 3-49). The deposition of

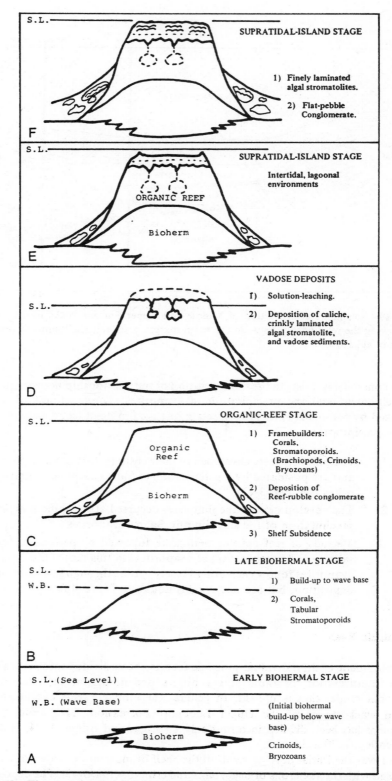

Figure 3-49 The postulated development of pinnacle reefs from the bioherm to the supratidal island stage (from Huh, Briggs, and Gill, 1977).

Example 8   163

these mounds began in quiet water below wave base. As the mounds grew they were colonized by corals and stromatoporoids and, finally, when they reached the high-energy wave zone, frame-building organisms became dominant and the mounds passed into their reefal stage.

Stage 2 of pinnacle growth was characterized by the development of a wave-resistant organic reef or boundstone (Fig. 3-49). The reef fauna included frame-building stromatoporoids, tabulate corals, sediment-binding algae, crinoids, bryozoans, and brachiopods. Changes in the dominant organisms in the reef assemblages occurred with reef growth, and there was a progression from stromatoporoids to corals and finally to algae. Reef growth kept pace with subsidence resulting in a vertical pinnacle-reef morphology. At the climax of growth the pinnacles stood from 90 to 180 meters above the seafloor.

Stage 3 of pinnacle development is interpreted as a period of subaerial exposure when sea level fell as estimated 9 to 15 meters. During this time the reef core underwent freshwater leaching with the development of vugs, internal sediment, vadose pisolites, calcrete, and laminar calcite.

During Stage 4 the top of the pinnacles formed supratidal islands characterized by the development of stromatolites, burrowed mudstone, peloidal wackestone, and flat-pebble conglomerate. During this stage the sea level fluctuated and the reefs alternated between the supratidal, intertidal, and subtidal zones (Fig. 3-49). With the development of hypersaline conditions and fluctuating sea levels the interreef areas were filled by carbonates and evaporites, and the reefs themselves were finally overlain by a thick layer of anhydrite.

Diagenesis and Reservoir Occurrence

The pinnacle reefs were affected by freshwater leaching, dolomitization, and the precipitation of void-filling evaporite minerals during their early history, which has had a great influence on subsequent porosity and permeability.

During periods of subaerial exposure, either prior to or during the formation of the Salina evaporites, the pinnacles had their porosity enhanced by freshwater leaching and the creation of vadose fractures, vug, and channel porosity in both the biohermal and organic reef facies. Dolomitization occurred in the postulated zone of freshwater/marine-water mixing, which enhanced porosity especially in the biohermal skeletal micrites (Petta, 1980). Porosity occlusion during exposure was the result of introduction of internal sediment and calcite cements (Bentley, 1979b).

During stages of evaporite precipitation the formation waters within the pinnacles became hypersaline, and available porosity was plugged by anhydrite and halite. A thick salt-plugged zone developed as sea level rose, and the tendency in some reefs is an increase in the amount of salt plugging toward the top of the reef (Petta, 1980).

Within producing reefs intercrystalline and vuggy porosity occurs in the biohermal and reef-core facies. Porosities range from 3 to 37% with an average of 6% (Gill, 1979).

Large-scale trends in diagenesis and production have been noted in the pinnacle-reef belt (Gill, 1979). The plugging of the reefs with halite and anhydrite appears to increase basinward, whereas the degree of dolomitization and the amount of secondary porosity increase toward the shore. The trend in dolomitization results in reefs that are composed completely of dolomite on the shoreward edge of the pinnacle belt, reefs of interbedded limestone and dolomite within the belt, and reefs completely composed of limestone on the basinward edge of the shelf.

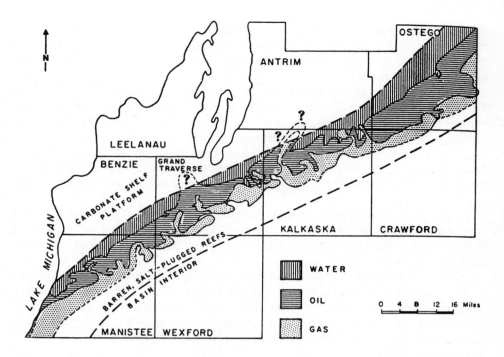

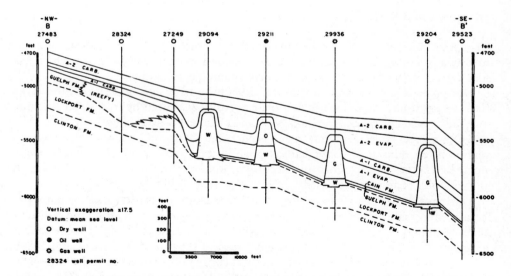

Figure 3-50 Map showing the distribution of salt-plugged, gas-, oil-, and water-bearing reefs in northern pinnacle-reef belt of the Michigan Basin. The cross section shows the segregation of fluids within the reefs (from Gill, 1979).

Example 9   **165**

The distribution of reservoir fluids is shown in both plan and section in Figure 3-50. On the basinward margin of the shelf the reefs tend to be barren or salt plugged. Moving in a shoreward direction the reefs produce gas, then oil, and finally closest to the shore they are barren or produce water. This distribution has been explained as the result of up-dip migration and differential entrapment of fluids in the reefs which are hydraulically connected through the underlying Lockport Formation (Gill, 1979).

Conclusions

The pinnacle reefs of the Michigan Basin form small, isolated petroleum reservoirs encased in impermeable evaporites. The temporal relationships between the reef sequence and the evaporites are still in dispute. Pinnacle development occurred in four stages and included periods of subaerial exposure which enhanced reef porosity and permeability through leaching and dolomitization. Subsequent evaporite precipitation filled much of this porosity, producing many reefs that are completely salt plugged and impermeable.

Regional trends have been recognized across the pinnacle-reef belt, and these predict increased salt-plugging of the reefs basinward, increased dolomitization and preserved secondary porosity toward the shore, and zones of production that pass shoreward from gas to oil and finally to water. The producing reefs have porosities that range from 3 to 37% and permeabilities of 11 to 12 md.

EXAMPLE 9.   CARBONATE RESERVOIRS IN A MARINE SHELF SEQUENCE, MISHRIF FORMATION, CRETACEOUS OF THE MIDDLE EAST

J. Reulet

The Mishrif reservoirs have been studied from several localities in the Middle East, especially from Iraq. They are Middle to Late Cenomanian in age and are composed of rudistid packstones and grainstones (Fig. 3-51). As a whole, the Mishrif is a regressive sequence overlying basinal limestones. It is composed of outer-shelf limestones overlain by inner-shelf limestones truncated by an unconformity. Evolution of the Mishrif sequence was progressive or occurred in fluctuating steps linked with instability of the shelf. An example of the latter type of evolution is given here.

Analysis of the Mishrif

The Mishrif Formation is made up of outer- and inner-shelf deposits separated by a barrier.

The outer-shelf deposits include basin facies of mudstone-wackestones with pelagic foraminifera and wackestone-packstones with *Oligostegina*. On the outer-shelf side of the barrier low- to moderate-energy wackestones with echinoderms, prealveolinids, and other bioclastic debris occur together with higher-energy algal pisolite packstone-grainstones containing orbitolinids, echinoderms, and green algae.

The barrier deposits include packstone-grainstones with rudistid debris. These deposits formed in a high-energy environment as thin biostromes rather than as real

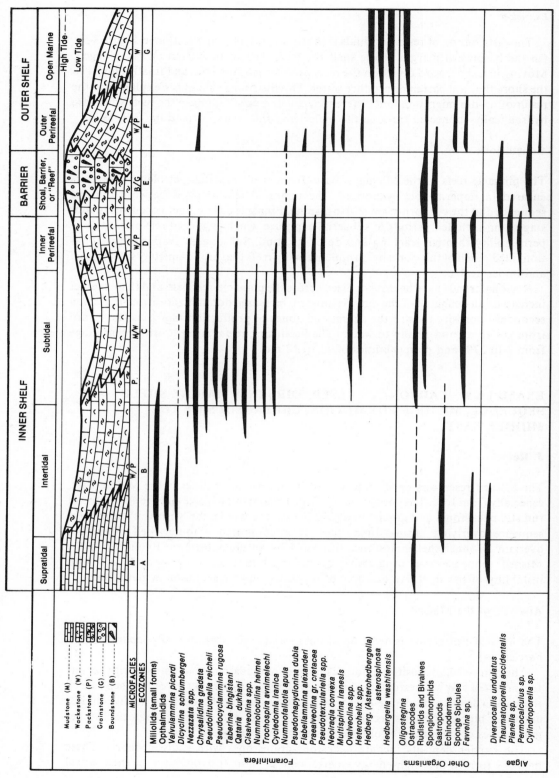

Figure 3-51 Distribution of principal organisms in the depositional environments of the

Example 9 167

reefs, although locally there are buildups that may be termed reefal. Whole rudistids in life position have not been found, and the coarsest debris present is only a few centimeters across.

The inner-shelf deposits include wackestone-packstones with rudistid debris, labeled inner perireefal on Figure 3-51. These are moderate- to high-energy deposits corresponding to back-"reef" buildups or small banks that occur within the lagoon. These deposits are from 1 to 1.5 m thick. Mudstone-wackestones with benthonic foraminifera (peneroplidae, miliolidae, alveolinidae, and others), green algae, gastropods, and sponge spicules are low-energy deposits formed in the subtidal zone behind the barrier. Peloidal wackestone-packstones with miliolids, rudistid debris, and gastropods are high-energy deposits formed in the intertidal zone. The inferred supratidal deposits are low-energy mudstones (Fig. 3-51).

The use of different logging techniques for the study of carbonate rocks is complementary to sedimentological analysis. The physical parameters measured by the logs allow continuous monitoring of lithology and other important parameters, such as porosity.

Comparison of different well logs with facies or depositional environments makes it possible to delineate formations within sedimentary sequences based on log characteristics.

In the Mishrif, where the formation is wholly of carbonate, the lithological subdivision is no problem, and the facies and environments can be characterized in a straightforward fashion. Based on log analysis, the following observations can be made:

1 High-energy deposits related to the barrier or shoal environments are characterized by significant porosity, which is visible in different porosity logs (neutron, density, and sonic logs and through weak radioactivity on a gamma-ray log—10 API maximum).

2 The shallow open marine environments are of low to moderate energy. They contain important clay seams and stylolites commonly associated with dolomite layers. These layers result in moderate radioactivity (between 15 and 5 API) associated with argillaceous material and organic matter. A fine intramicrite porosity produces an irregular curve on the logs, indicating porosity of 8 to 16%, which is not to be overlooked, but is nevertheless lower than the porosities found in the high-energy deposits.

3 The restricted marine environment situated behind the barrier shows various features linked with the bathymetry and energy of the inner shelf. The most common deposits appear to be of low energy and subtidal. They are rich in micrite, clay seams, and stylolites and have a radioactivity varying from 15 to 30 API, reaching as high as 50 API. They have very low porosity. Within the subtidal deposits thin rudistid banks can be distinguished from the surrounding limestone by their significantly higher porosities, usually greater than 15%.

These examples show that the use of well-logging techniques in the Mishrif, such as radioactivity and porosity logs, makes possible quick and continuous recognition of the major facies and depositional environments. The results are simplified because slight lithological variations, multiple diagenetic effects, and variations in faunal and sedimentological characteristics cannot be differentiated by well logs. The technique

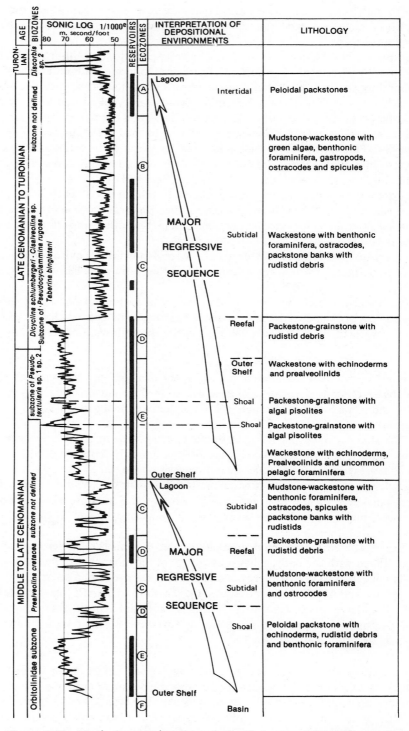

Figure 3-52 Vertical succession through the Cretaceous Mishrif Formation.

Example 9   169

has a low degree of resolution and measures physical parameters only. Some of the problems with well logging, notably the origin and significance of radioactivity, diagenetic phenomena, and their relationship to log responses, are fundamental to carbonate studies, but are not as yet fully understood.

Vertical- and Lateral-Sequence Development

The Mishrif Formation is a regressive sequence underlain by basinal facies and capped by a regional unconformity. The Mishrif can be subdivided into two sequences both of which are regressive (Fig. 3-52): (1) the basal sequence, which ranges from open marine basin to restricted lagoonal, and (2) the upper sequence composed of outer-shelf deposits overlain by inner-shelf deposits. A sedimentary discontinuity apparently separates the two sequences.

Each of the two sequences can be further subdivided into a number of subsequences, including shoal, barrier, and lagoonal sequences (Fig. 3-53). The shoal sequences are composed of wackestones with echinoderms and alveolinids at the base and algal pisolite packstone-grainstones at the top. The barrier sequence from base to top consists of wackestones with echinoderms and alveolinids, packstones with fine to medium rudistid debris, echinoderms and algal pisolites, and grainstones with coarse rudistid debris and sparse corals. The lagoonal sequences are composed of mudstone-wackestones with benthonic foraminifera, green algae, ostracodes, and sponge spicules overlain by skeletal concentrates of fine to medium rudistid debris.

The Mishrif Formation is laterally continuous over large areas of the Middle East

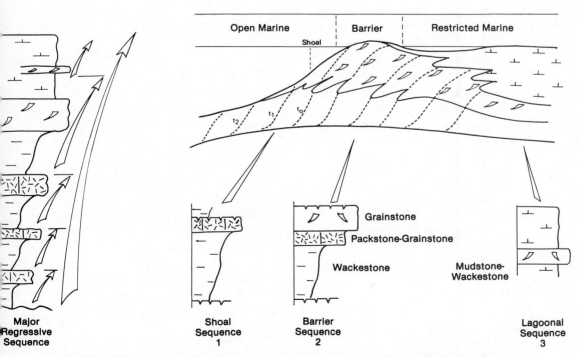

Figure 3-53 Interpretation of different depositional environments in the Mishrif Formation, Cretaceous of the Middle East.

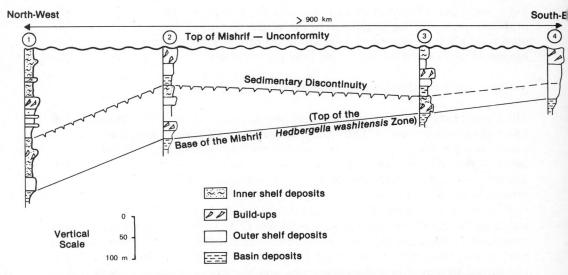

Figure 3-54   Facies changes in the Mishrif Formation, Cretaceous of the Middle East.

(Fig. 3-54). The two major regressive cycles are large-scale features, but in detail they vary in lithology and thickness.

Section 1 (Fig. 3-54), taken as an example of the Mishrif Formation, is basically a marine shelf sequence. A discontinuity within the sequence is indicated by the superposition of outer-shelf deposits on inner-shelf deposits. Sections 2 and 3 show basinal deposits at the base of the upper sequence overlying restricted marine deposits.

Mishrif Reservoirs (Figs. 3-55 and 3-56)

The best reservoirs occur in the high-energy barrier and shoal deposits (Fig. 3-55). Less important reservoirs occur in low-energy facies, such as the outer- and inner-shelf deposits. Average porosity and permeability values for the various facies are listed below.

Barrier facies ("reefal")	Porosity 15 to 20%
	Permeability 10 to 100 md
Shoal facies	Porosity 20 to 25%
	Permeability can exceed 1000 md
Outer-shelf facies (moderate energy)	Porosity 10 to 15%,
	Permeability ± 10 md
Inner-shelf, facies (moderate energy)	Porosity 10 to 15%,
	Permeability 0 to 10 md
(low energy)	Porosity 0 to 5%,
	Permeability 0.1 md

However, variations in sedimentary conditions and diagenetic alteration can significantly alter these values. The reservoir characteristics therefore depend upon:

1 *Sedimentological conditions.* Of particular importance in high-energy deposits is a lack of micrite together with a well-developed primary porosity and good

Example 9　　171

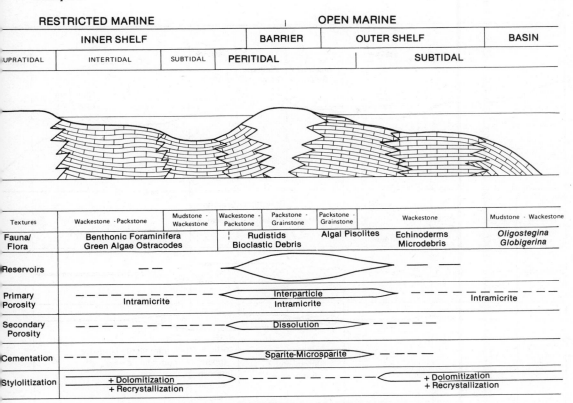

RESTRICTED MARINE				OPEN MARINE		
INNER SHELF			BARRIER	OUTER SHELF		BASIN
SUPRATIDAL	INTERTIDAL	SUBTIDAL	PERITIDAL	SUBTIDAL		

Textures	Wackestone - Packstone		Mudstone - Wackestone	Wackestone - Packstone	Packstone - Grainstone	Packstone - Grainstone	Wackestone		Mudstone - Wackestone
Fauna/Flora	Benthonic Foraminifera Green Algae Ostracodes			Rudistids Bioclastic Debris		Algal Pisolites	Echinoderms Microdebris		*Oligostegina Globigerina*
Reservoirs									
Primary Porosity	Intramicrite				Interparticle Intramicrite			Intramicrite	
Secondary Porosity					Dissolution				
Cementation					Sparite-Microsparite				
Stylolitization	+ Dolomitization + Recrystallization						+ Dolomitization + Recrystallization		

Figure 3-55　Reservoir characteristics of the Mishrif Formation.

pore interconnections. Porosity tends to be interparticle and intraskeletal (the latter may show significant water saturation despite the presence of oil). The moderate-energy facies of the fore-barrier and shallow open marine deposits which contain abundant micrite have important intramicrite porosity consisting of fine pores with thin interconnections.

2 *Diagenetic conditions.* These can either improve or destroy porosity. Dissolution is particularly prominent in the barrier facies (Fig. 3-55), where it improves interparticle porosity and locally creates large voids several centimeters across. Many of the solution features are related to the unconformity surface at the top of the formation. However, this interval is uncored and the cuttings are contaminated by the overlying clays, so that the dissolution structures are difficult to document.

Fracturing is of two main kinds. Fractures 10 to 20 cm long of tectonic origin can be seen in the core samples. These are normally subvertical and either healed or open, usually less than $\frac{1}{10}$ mm wide. The intensity of fracturing (length of fracture per meter of core) has been measured and compared between wells. It varies greatly from structure to structure and even between wells on the same structure.

Microfractures, from 1 to 3 cm long, link horizontal or vertical stylolite seams. These fractures are very common in the micrite-rich layers (wackestones) that are conducive to stylolitization. Fractures are almost absent in layers devoid of micrite and in areas of coarse bioclastic debris.

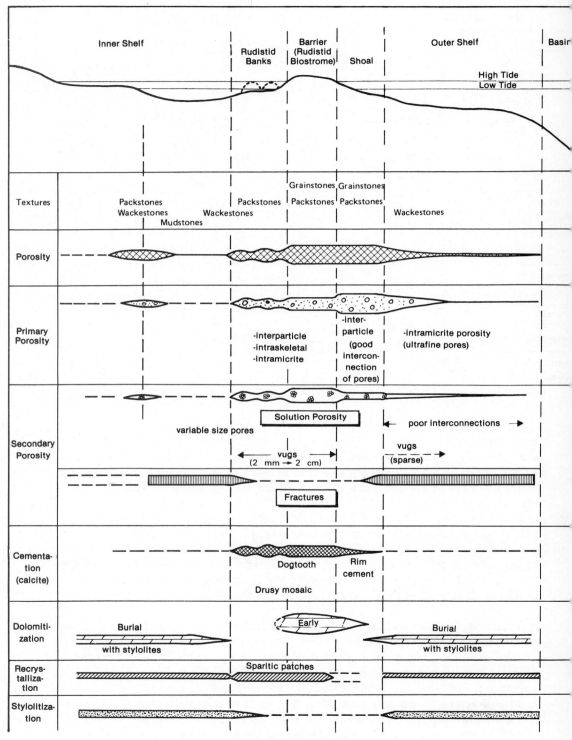

Figure 3-56 Characteristics of the Mishrif reservoirs.

172

Example 10   173

Cement, in the form of drusy mosaics and dogtooth spar, is present in the barrier buildups. The cement is not widely developed and has only destroyed part of the interparticle porosity. The sedimentological and diagenetic events included the invasion of certain of the structures by freshwater and accompanying cementation.

Stylolites form in layers rich in micrite (mudstone-wackestones). Diffuse "horse-tail"-stylolite seams occur and merge to form subhorizontal stylolites. Dolomite rhombs are found concentrated along the stylolites, creating brown patches in the rock. In stylolite zones local recrystallization of micrite occurs, destroying the intramicrite porosity and forming impermeable white nodules.

Conclusions

The Mishrif Formation is a multibedded reservoir composed entirely of carbonates. The reservoir zones, which are of variable quality, are separated by dense, nonporous beds. The rudistid barrier buildups and the bioclastic shoals form the best reservoirs.

The diagenetic processes favorable to the development of porosity and permeability (dissolution and fracturing) are generally more common than those that destroy a potential reservoir (cementation and pressure solution). However, in places diagenetic changes put reservoir facies in zones where destructive processes are of increased importance.

EXAMPLE 10. BARRIER-REEF AND CARBONATE-SHELF SEDIMENTATION: LENNARD SHELF, DEVONIAN OF WESTERN AUSTRALIA

R. Elloy

The Lennard Shelf forms the northern edge of the Canning Basin in the north of Western Australia (Fig. 3-57), overlying Precambrian basement rocks of the Kimberley block. The carbonate rocks of the shelf are mainly Devonian in age and crop out along a 300 km belt oriented northwest-southeast. These rocks provide a spectacular example of an exposed Paleozoic barrier-reef complex. The Devonian carbonates are transgressive over the Precambrian basement and locally over Ordovician rocks. Carboniferous siltstones and Permian sandstones of the Fairfield and Grant Formations overlie them. In the subsurface, rocks of the Lennard Shelf are of prime interest for oil exploration.

Geological Setting

The Devonian rocks of the Lennard Shelf crop out along a series of ranges several tens of kilometers long and oriented northwest-southeast over a distance of 300 km. The ranges have abrupt relief, but do not vary in height by more than 1000 m. Most of the outcrops are of shelf facies which abut the Ordovician and Precambrian highs. The slope facies flank the ranges at irregular levels and grade outward into deep-basin lithologies. These basin deposits form low areas covered with black soil.

The shelf area is almost flat, with a slight tilt to the south. The dominant faults are parallel to the ranges, particularly the Pinnacle Fault, which occurs to the south of the

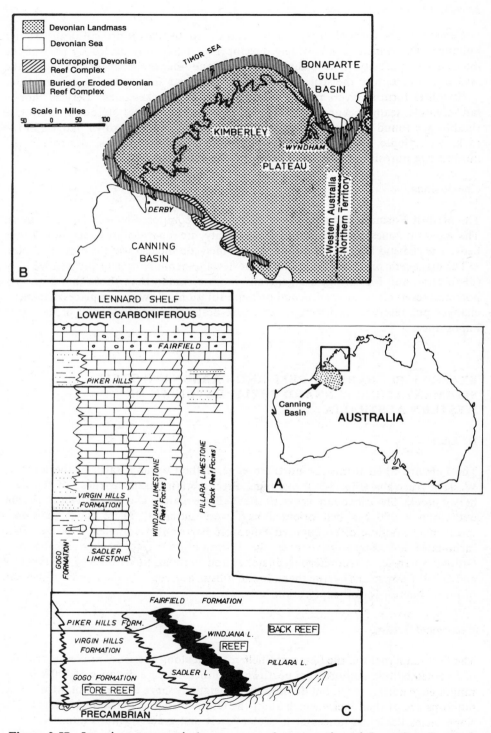

Figure 3-57 Location map, vertical sequence, and cross section of the carbonate complex of the Lennard Shelf, Devonian of Western Australia. (*A*) Index map of Australia. (*B*) Map of the northern Canning Basin showing the position of the Devonian reef complex (after Playford, 1980). (*C*) Stratigraphic column and cross section through the Lennard Shelf.

Example 10   175

ranges. To the southeast the major structural trend is north-south, which influences the distribution of the sedimentary bodies.

The carbonate shelf deposits are Middle to Late Devonian (Givetian to Famennian) in age. They include inner-shelf, reef, and open marine deposits. The inner-shelf and reef deposits contain only facies fossils and are thus rarely datable. The age of the slope and basin facies can be given with more accuracy, since these contain abundant brachiopods, nautiloids, and conodonts. The stratigraphic terminology of Playford and Lowry (1966) has been used in this study.

Reef and inner shelf facies	Windjana Lst.	
	Pillara Lst.	Givetian
Slope and basin facies	Gogo, Sadler,	to
(including off-reef and	Napier, Virgin	Famennian
interreef)	Hills and Piker	
	Hills Fms.	

Sedimentary Bodies

A special feature of the Lennard Shelf sequence is the preservation of the original shape of the sedimentary bodies. The present relief is the result of the combined effects of tectonism and erosion, but these processes have not obliterated original depositional slopes and relief. The depositional dips are shown by geopetal fabrics and in places are as high as 30°. Slumps and sedimentary breccias are common, and algal stromatolites are seen to have grown vertically on dipping strata. Distinct lithologies mark the passage from high-energy reefal deposits into low-energy open marine sediments.

The reefal bodies have two main forms. Tabular forms occur abutting the uplifted Precambrian, whereas regular forms occur encircling atolls around basement highs. The tabular bodies are of two kinds—those that consist of frame-building organisms forming true biostromes and flat bioaccumulations composed of bioclastic debris (peloids, oncolites, oöids, and biogenic debris) with subordinate frame-builders. The difference between the two kinds is not clear cut, since they tend to pass gradationally into each other.

A typical tabular biostrome from the Pillara Range is composed of peloidal micrite with *Amphipora* alternating with layers of limestone containing spherical stromatoporoids, isolated corals, and *Thamnopora*. Intercalated with these are layers having bird's eye structures and containing large gastropods.

The vertical lithological succession shows an increase in the abundance of frame-building organisms: *Thamnopora, Hexagonaria,* stromatoporoids, and sticklike rugose corals. The uppermost part of the section overlying an angular discordance contains grey-green brachiopod-crinoid coquinas with breccias and concretions.

The flat bioaccumulations have a similar shape to the biostromes, and a distinction between the two is made only on the organisms present. The best examples of bioaccumulations are found in the Oscar Range (Fig. 3-58). The lithologies present are similar to those found in the biostromes, including peloidal micrite, containing *Amphipora*. Oöids also occur and are well sorted or may show graded bedding and channeling. The oöids may themselves cut channels into underlying peloidal mudstone. The oölitic deposits have in places migrated downslope to intertongue with

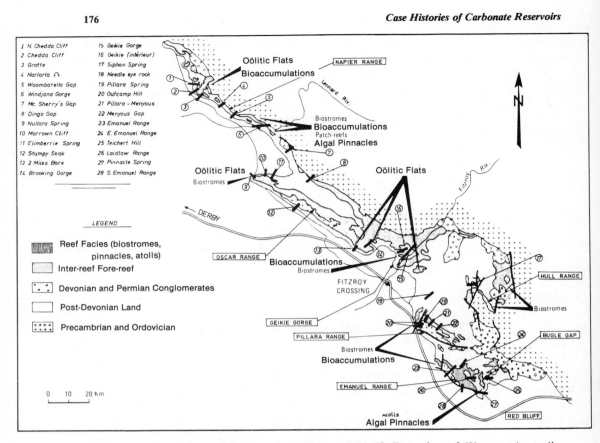

Figure 3-58 Location of field sections, Lennard Shelf, Devonian of Western Australia. Map after Playford and Lowry, 1966.

the brachiopod-crinoid coquinas. Oncolites may occur associated with the oöids, and algal buildups or small algal reefs may occur on the surface flats of the mounds. The slopes of the bioaccumulations are characterized by the same sediments as occur around the biostromes, such as mudstones with brachiopods, crinoids, concretions, breccias, and layers with pelagic fauna (*Orthoceras* and goniatites).

Circular bioherms can be of all sizes, ranging from small buildups to atolls that are measured in tens of miles.

Facies and Depositional Environments

The sedimentary rocks of the Lennard Shelf contain a wide range of organisms and sedimentary structures. This is not surprising, since the shelf edge would have been situated between the paleoequator and 3° S during the Middle to Late Devonian. These are the latitudes where organisms are very abundant and diverse, especially within a carbonate sedimentary environment. The following features occur in the facies of the Lennard Shelf and are classified as VA — very abundant, A — abundant, QA — quite abundant, and R — rare.

Example 10   177

1   **Organisms**
  (a)   Frame-builders
          Stromatoporoids (VA)
              Amphipora (VA)
              Stachyodes (A)
          Rugose corals
              Dysphyllum (A)
              Themnophyllum (A)
              Hexagonaria (A)
              Phillipsastrea (R)
          Tabulate corals
              Alveolites (R)
              Thamnopora (R)
  (b)   Encrusting organisms
          Sphaerocodium (VA)
          Renalcis (VA)
          Girvanella (R)
          Columnar stromatolites (R)
          Laminar stromatolites (R)
  (c)   Other organisms
          Crinoids (VA)
          Brachiopods (VA)
          Gastropods (VA)
          Molluscs (A)—*Megalodon* (R)
          Receptaculites (QA)
          Nautiloids (VA)
          Calcispheres—*Iregularina* (R)
          Trilobites (R)
          Foraminifera (R)
          Bryozoans (R)

2   **Nonorganic constituents**
  (a)   Peloids (VA)
  (b)   Lumps (A)
  (c)   Oöids (A)
  (d)   Oncolites (QA)
  (e)   Detrital quartz and rock fragments (QA)

3   **Sedimentary structures**
  (a)   Slope breccias and slumps
  (b)   Directional indicators (oriented crinoid stems and nautiloids)
  (c)   Geopetal structures, mainly within brachiopods
  (d)   Desiccation structures

   (e) Diagenetic structures (stromatactis, concretions, internal sediment, and neptunian dikes)

   (f) Channels

4 Color. Colors vary from white through yellow, green, and red. They are rarely very dark. The different colors may be significant environmental indicators

The components listed in the four categories above are shown in relation to a profile across the Lennard Shelf in Figure 3-59. A comparison of the different sedimentary rocks and their components has led to the definition of 12 facies (Table 3-6), which are the products of three principal environments:

1 Shelf, including back-reef or lagoon, high-energy outer-shelf and buildup facies.

2 Shelf-to-slope transition.

3 Slope and deep-basin environment.

Some of the facies can be divided into subfacies where the abundance of certain constituents varies.

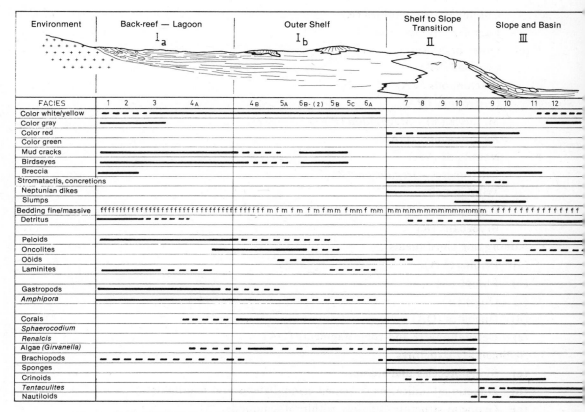

Figure 3-59 Distribution of environmentally controlled features along a morphological profile adapted to the Lennard Shelf.

Example 10 179

Table 3-6 Relationship between Facies and Depositional Environments on the Lennard Shelf, Devonian, Western Australia

	Facies	Environment		Comments	
1	Conglomerate with basement rock pebbles, rich in quartz, in places dolomitic, contains large gastropods, is red to brown and green in color Mud cracks	Inner	Shelf	Lagoon or back-reef, mostly peloidal mudstones	Ia
2	Laminites with desiccation cracks			Mudstones and wackestones passing into boundstones with *Amphipora*, stromatoporoids Low energy	
3	Peloidal mudstone with gastropods, birdseyes, rare *Amphipora*				
4	Peloidal mudstones with 4 A — *Amphipora* dominant 4 B — *Amphipora* and other stromatoporoids				

5	Calcarenites
	  5 A — with oncolites
	  5 B — with oöids
	  5 C — with a variety of fragments
		Outer		Calcarenites and boundstones. High energy.
				
				In certain areas on the shelf edge
				build-ups occur, also bioaccumulations and oöids.
					Ib

6	Build-ups
	  6 A — with corals and stromatoporoids
	  6 B — with algae				

7	Argillaceous limestone and mudstone with brachiopods, rare crinoids, in places a coquina. Mainly green in color.	Shelf to Slope Transition		Bafflestones Low energy	II
8	Varicolored mudstone with concretions, *Renalcis, Sphaerocodium*, sponges, stromatactis. Internal sediment				
9	Breccias, megabreccias			Slope currents High energy	
10	Calcarenites with crinoids, commonly red				
11	Mudstones and wackestones with pelagic organisms large nautiloids and pteropods	Basin		Low energy	III
12	Peloidal mudstones with fish remains, pelagic fauna and rare oncolites				

Shelf Environment

The back-reef and lagoonal facies of the inner shelf include facies 1 to 4 (Table 3-6). Facies 1 is the basal facies of the sequence and is transgressive across the Precambrian or Ordovician basement. It is a conglomerate composed of pebbles of basement rock from many different sources in a matrix of peloidal sand. This matrix is commonly dolomitized. Bird's-eye structures are abundant, and the fauna consists of large gastropods, uncommon brachiopods, and nautiloids. The color of the rocks varies from brown through red and green.

Facies 2 consists of laminites associated with desiccation structures, whereas facies 3 is a peloidal mudstone with massive bedding and mudcracks.

Facies 4 consists of peloidal mudstone containing *Amphipora*. A variation on this occurs where the *Amphipora* are very abundant and form a boundstone (facies 4a). Bird's-eye structures are also found in these rocks. Facies 4b is light in color and contains stromatoporoids, including *Amphipora* and *Stachyodes,* rugose and tabulate corals, and uncommon gastropods. The basal part of facies 4b is peloidal.

Facies 5 and 6 constitute the outer-shelf deposits. Facies 5 consists of bioaccumulations, with oncolites (5a), oöids (5b), and calcarenites (5c). The rock framework is locally peloidal. In the example containing oöids, the oöids form a grainstone with graded bedding, channels, and pockets of calcarenite composed of a variety of fragments. Nautiloids that originate in the basin have been washed in and deposited in the shelf sediments. Bird's-eye structures are present, indicating periods of emergence.

Facies 6 is the biostrome or buildup facies. Facies 6a consists of coral buildups with dominant stromatoporoids, either rounded or branching forms. Sporadic *Hexagonaria*, the tabulates *Alveolites* and *Thamnopora*, and the rugose coral *Dysphyllum* also occur. Facies 6b is composed of algal buildups, which occur in the Bugle Gap and Elimberrie Spring regions (Fig. 3-58).

Shelf-to-Slope Transition

This part of the sequence includes facies 7 to 10 inclusive. Facies 7 is a greenish argillaceous limestone containing brachiopods. In places it is a true coquina with a single species of brachiopod, but more commonly it includes both atrypids and spiriferids, which contain geopetal sediment. This sediment has a uniform orientation and indicates deposition on a slope.

Facies 8 is a red to bright green concretionary mudstone. It includes both frame-building and pelagic constituents, indicating some communication with the basin.

Facies 9 is a breccia, in places composed of blocks several cubic meters in size and derived from the two preceding facies. The breccia has a calcarenitic matrix composed of crinoidal, coral, and shell fragments.

Facies 10 is a red, pink, or white calcarenite composed of crinoids, shell fragments, and fragmented corals. It represents part of the gradation from a breccia into the peloidal sediments of the basin. The imbrication of crinoid stems in facies 10 can be used to indicate current direction on the sediment slope.

The facies described from the shelf-to-slope transition usually occur in sequence from 7 at the base to 10 at the top. However, this is not ubiquitous and sequences may have one or several of the facies absent or occurring as lateral equivalents.

Slope and Deep-Basin Environment

The facies that represent these environments are poorly exposed because they occur at the base of modern slopes and form a plain covered with black soil.

Facies 11 consists of dark mudstones and wackestones with pelagic organisms such as goniatites, *Orthoceras,* and *Tentaculites.*

Facies 12 consists of peloidal mudstones. In the Needle Eye Rock region these

Example 10   181

sediments are rich in quartz—10% on the average, but in places up to 40%. Oncolites are uncommon, and crinoids, sponge spicules, and radiolarians are present. In the field sections east of Emanuel Range the light yellow peloidal mudstones are dominant and contain nodules with rare fish fossils. This facies passes into the shelf-to-slope transition in the Emanuel Range, Bugle Gap, Teichert Hill, and Outcamp Hill region (Fig. 3-58).

The boron content of the rocks of facies 12 averages 80 ppm, with a maximum concentration of 130 ppm. The average barium content is 300 ppm, whereas vanadium has an average concentration of 105 ppm. Some enrichment in zinc occurs—600 to 900 ppm, with an average of 180 ppm. The basin facies are equally impure, with insoluble residues averaging 16%.

Table 3-6 summarizes the different depositional environments and corresponding facies together with their principal characteristics. Figures 3-60 and 3-61 show facies and depositional environments across the major field sections (Fig. 3-58).

In some sections a particular environment may be completely absent as in Geikie Gorge, where the inner shelf passes directly into the slope facies; in others the oölitic facies is poorly developed on the outer shelf, suggesting the lateral development of a barrier.

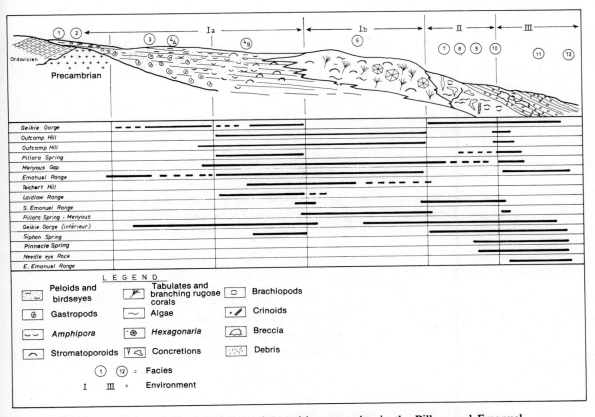

Figure 3-60 Facies and environments of deposition occurring in the Pillara and Emanuel Ranges, and in the Geikie Gorge sections, Western Australia.

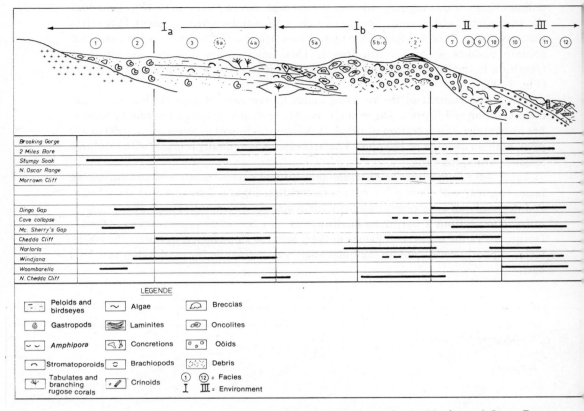

Figure 3-61 Facies and environments of deposition found in the Napier and Oscar Range sections, Western Australia.

Relationships Between Facies: The Sequences

Inner shelf biostromes are typified by sequences containing facies 1 through 4, although facies 1 and 2 are commonly absent. Three typical biostrome sequences are given below.

Outcamp Hill Sequence

	TOP
4 units	Peloidal mudstones with bird's-eyes and stylolites
4 m thick	Calcarenites with *Amphipora*
	Calcarenites with stromatoporoids
	Peloidal mudstones with branching corals

Limestone Billy Hill

	TOP
2 units	Peloidal mudstones with bird's-eyes
2 m thick	Calcarenites with *Amphipora*

Example 10 183

Pillara Spring

	TOP
	Peloidal mudstones
3 units	Peloidal mudstones with *Amphipora*
3 m thick	Calcarenites with stromatoporoids

Typical biostrome units are repeated numerous times on the inner shelf. Facies 3, 4a, and 4b are most commonly developed, but in the Emanuel Range facies 1 and 2 are also included. The biostrome sequences are typically shallowing-upward and are of economic significance as the hosts of all inner-shelf galena deposits.

The shelf-to-slope transition is typified by the following sequence:

TOP

Calcarenites with crinoids

Breccias and calcarenitic breccias

Peloidal mudstones with concretions and *Renalcis*

Brachiopod coquina

This sequence is composed of facies 7, 8, 9, and 10, it is not normally repetitive, and its thickness is measured in tens of meters.

Biostrome sequences containing framework-building corals and stromatoporoids are typified by the following sequence: facies 3 and 6 at the top, facies 3, 4a, or 4b in the middle, and facies 1 and 2 at the base. This is an evolution from an inner-shelf to outer-shelf environment, with a framework facies at the top. A similar sequence is shown by virtually all the biostromes.

The bioaccumulations show the same evolution as the biostromes in the inner shelf, but facies 5 occurs in the outer-shelf part of the sequence. The framework builders are replaced by oölitic barriers.

The Lennard Shelf Sequence (Figs. 3-62 and 3-63)

The most classical and complete sequence is found in the Emanuel Range section. From the transgressive facies of the inner-shelf deposition continued with outer shelf, shelf-to-slope transition, and finally open basin deposits. This deepening-upward sequence that indicates transgression reflects the evolution of the whole Lennard Shelf, with some exceptions. This sequence is of Type A, as shown in Figure 3-64.

A second kind of sequence, Type B (Fig. 3-64), which has the same general trend, occurs in Geikie Gorge and Brooking Gorge. In sequence B the buildup facies with corals and stromatoporoids is replaced by an oölitic or calcarenitic barrier.

At Dingo Gap, Pinnacle Spring, and Syphon Spring a third kind of sequence, Type C, occurs (Fig. 3-64). This sequence is characterized by the telescoping of the marine and lagoonal deposits with almost total lack of an intermediate barrier or buildup facies. This kind of sequence is best illustrated by examples from McSherry's Gap, Woombarella Gap, Bugle Gap, and Pinnacle Spring. Minor oöids occur in the last section. There is the indistinct development of a barrier at Slumpy Soak and a better example at Syphon Spring and Dingo Gap. Along the range the sequences developed

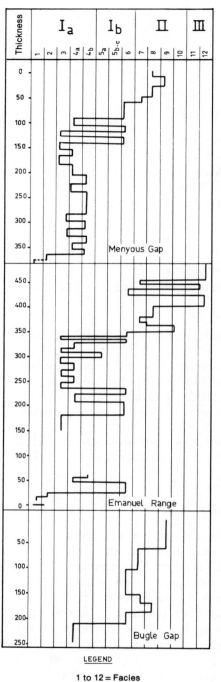

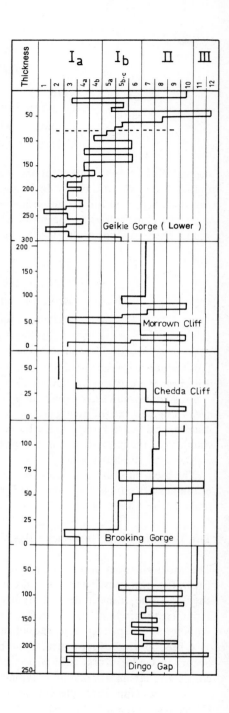

LEGEND

1 to 12 = Facies

I_a Inner Shelf

I_b Outer Shelf

II = Shelf to Slope Transition III = Slope and Basin

Figure 3-62 Sequences of environments seen in a number of sections, Devonian of the Lennard Shelf.

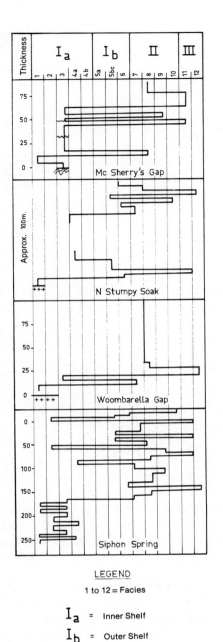

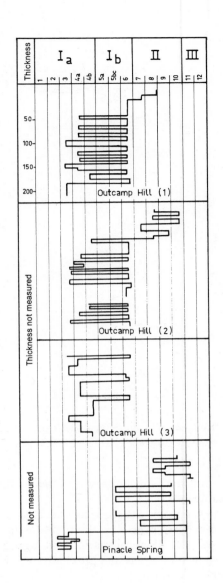

LEGEND

1 to 12 = Facies

I_a = Inner Shelf

I_b = Outer Shelf

II = Shelf to Slope Transition

III = Slope and Basin

Figure 3-63 Sequences of environments in a number of sections through the Devonian sequence on the Lennard Shelf, Australia.

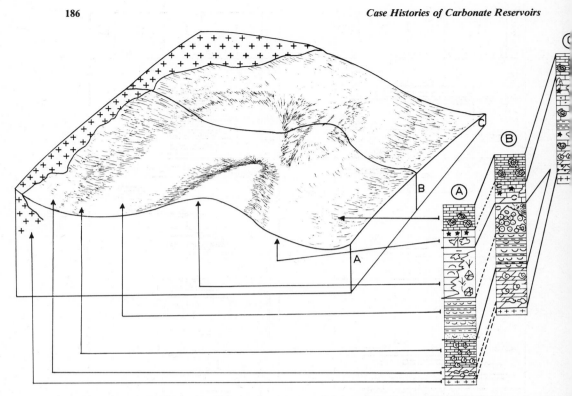

Figure 3-64 Three principal kinds of sequences developed on the Lennard Shelf, Devonian of Western Australia.

are not identical, but are of Type A, B, or C, all of which have the same general significance.

In summary, the sequences developed can be categorized as Type A, with a reefal-type barrier, Type B, with an oölitic or bioaccumulation barrier, or Type C, which shows telescoping of marine and lagoonal facies. Variations occur where no slope developed or where the proximity of a barrier is indicated by the presence of minor oöids or corals.

The Type A and B sequences show the least variation with little oscillation between adjacent facies. The Type C sequence implies a locally unstable zone, either an area of relief or of rapid marine incursion, which alternates with a lagoonal setting.

Along the Lennard Shelf the different faunal assemblages and kinds of sequences found enable a zonation to be made across the shelf (Fig. 3-65). From northwest to southeast:

1 *Napier-Oscar Ranges.* Marine facies overlapping inner-shelf facies, Type C sequences, and variants.
2 *Oscar Range, Geikie Range, Hull Range.* Oöid bars, oncolites, peloids, Type B sequences.
3 *Outcamp Hill, Pillara Range, Emanuel Range, parts of Bugle Gap.* Biostromes, Type A sequences.
4 *Bugle Gap region.* Development of bioherms, Type A sequences.

Example 10 187

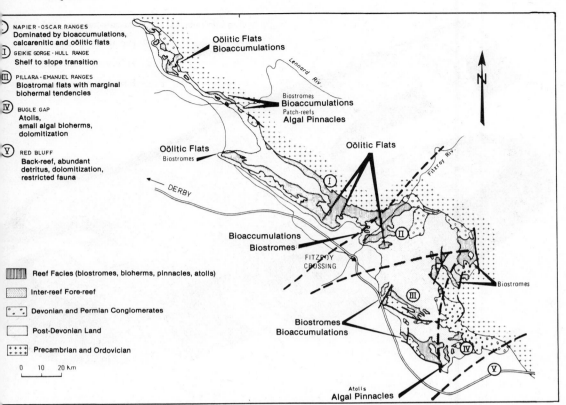

Legend:

Ⅰ NAPIER-OSCAR RANGES
Dominated by bioaccumulations, calcarenitic and oölitic flats

Ⅱ GEIKIE GORGE-HULL RANGE
Shelf to slope transition

Ⅲ PILLARA-EMANUEL RANGES
Biostromal flats with marginal biohermal tendencies

Ⅳ BUGLE GAP
Atolls, small algal bioherms, dolomitization

Ⅴ RED BLUFF
Back-reef, abundant detritus, dolomitization, restricted fauna

Reef Facies (biostromes, bioherms, pinnacles, atolls)

Inter-reef Fore-reef

Devonian and Permian Conglomerates

Post-Devonian Land

Precambrian and Ordovician

0 10 20 Km

Figure 3-65 Facies evolution on the Lennard Shelf (geological map after Playford and Lowry, 1966).

5 *Red Bluff region.* Outcrops are uncommon and the Upper Devonian basin-floor deposits are poorly exposed. The shallow marine sediments that outcrop are bright-red sandy micrites containing large gastropods.

A trend is indicated in a northwest-southeast direction (Fig. 3-65). To the north neither reefs nor oöids have developed. To the south oöids appear first and conditions gradually become more favorable for biostromal buildups until finally even atolls develop.

Details of the sequence in the Oscar Range are quite complex (Fig. 3-66). Oöid buildups occur preferentially on the southwest edge and on the north and south ends. Elsewhere micrites with gastropods and peloids with or without *Amphipora* are dominant. The diagrammatic representation of the region is to scale and is very similar to the diagram showing oöid distribution on the Bahama Bank (Fig. 3-67). The ooids in the Bahamas are concentrated along the borders of the bank and on the edge of the trench.

Biostromes with built-up edges compose the rocks of Pillara and Emanuel Ranges (Fig. 3-68). The buildups are formed of thin, uniform sequences that are commonly repeated. The sequences are shoaling-upward cycles in comparison to the overall transgressive megasequence that constitutes the Lennard Shelf. This situation is

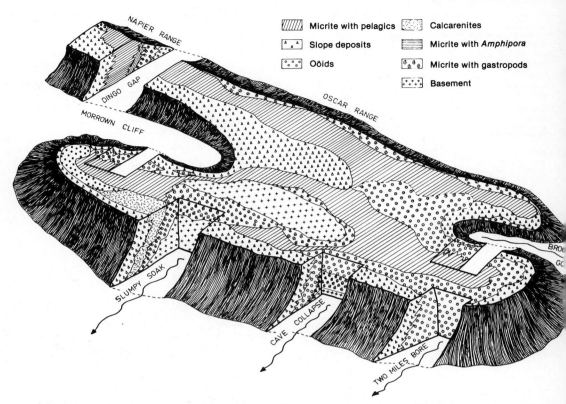

Figure 3-66 Theoretical block diagram of the Oscar and Napier Ranges, Western Australia.

similar but in the opposite sense to the sequences described from the cyclothems near the Village of Lofer in the Austrian Alps (Fischer, 1964). In the Alps the sequence varies from subtidal carbonate rocks with *Megalodon*, through intertidal laminites, to supratidal deposits. Supratidal deposits are rarely seen in the Pillara and Emanuel Ranges. Associated features found in the Alpine Triassic, such as neptunian dikes and brachiopod concentrations, are similar to the concretionary facies and the brachiopod coquina found in the sequence of the Pillara and Emanuel Ranges.

In modern sediments a similar sequence can be seen in Florida Bay. Here repeated short sequences occur composed of lacustrine deposits, mud mounds or sand bars (subtidal), beach and levee deposits with laminites and bird's-eye structures (intertidal), and dolomite crusts perforated by mangroves and interlayered with peat (supratidal). The sequence in Florida Bay shows more striking evidence of emergence.

Conclusions

The sequences that occur on the northern edge of the Canning Basin more closely resemble those of a complex basin edge than a simple barrier assemblage. Both bioaccumulations and reefs occur but the latter are subordinate. The dominant constituents of the sequence are interpreted as either biostromal, oölitic, or calcarenitic flats. Synsedimentary slopes developed during the growth of the biostromes and reefs

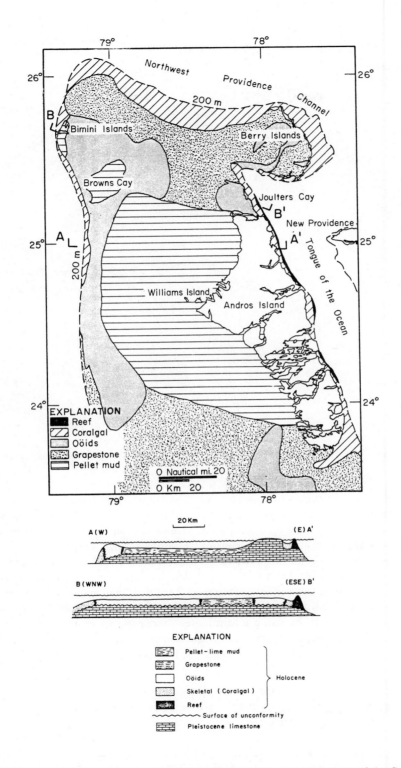

Figure 3-67 Map and profiles showing sediment facies on the northern part of the Great Bahama Bank (from Friedman and Sanders, 1978, p. 372, 373, modified from N. D. Newell, Imbrie, and others, 1959, p. 199, and E. G. Purdy, 1963, p. 473).

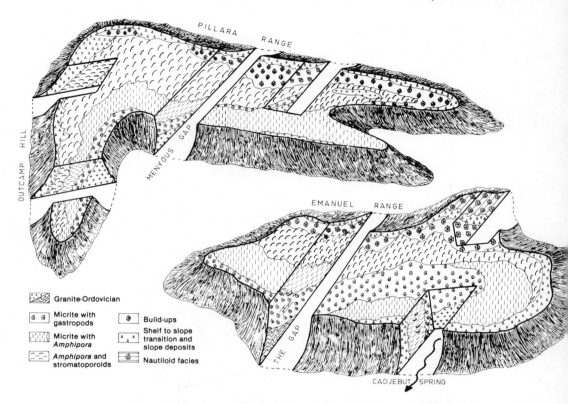

Figure 3-68   Theoretical block diagram of the Pillara and Emanuel Ranges, Western Australia.

proper. The interpretation of this sequence based on a single model from a modern sedimentary setting, such as the Bahamas, Florida, Persian Gulf, or Australian Great Barrier Reef, may be tempting but leads to extreme simplification. The Lennard Shelf is a complex sequence and as such cannot be compared to a single model. However, several parts of Holocene models can be used. These include the Australian Great Barrier Reef, Florida, and the Bahamas. Paleozoic and Mesozoic sedimentary rock sequences can also be used in interpreting the Lennard Shelf sequence. Wilson (1975) offers one of the best approaches. This consists of comparing ancient and modern models, adapting them by using additional information from detailed studies, and finally reaching a close approximation to the sequence in question. Every individual study has its own particular features and the Lennard Shelf is no exception.

EXAMPLE 11.   THE PERMIAN OF WEST TEXAS AND NEW MEXICO, U.S.A.; MULTIPLE RESERVOIRS IN REEF, SHELF, AND BASIN SETTINGS

The Midland Basin and Delaware Basin of West Texas and New Mexico are structural depressions separated by the north-south trending horst of the Central Basin Platform (Fig. 3-69). During the Permian the Delaware Basin rapidly subsided to become deep and stagnant with a partial filling of fine sands and silts (King, 1948; Newell et al.,

Example 11   191

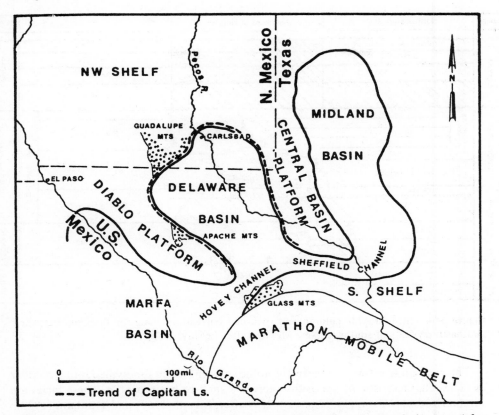

Figure 3-69 Location of the Delaware Basin and surrounding structural elements (after King, 1948, and Babcock, 1977).

1953). The deep basin rim was encircled by carbonate buildups, which have been interpreted as both classic barrier reefs and as marginal carbonate mounds. These buildups, with their adjacent slope deposits, separate the deep basin from the shallow, quiet environment of the back-reef shelf area, a site of carbonate, evaporite, and red bed deposition.

Oil production from the Permian carbonates occurs primarily from strata behind the shelf margin, but also from reef and deep-water talus deposits. The subsurface discrimination of shelf and basin deposits is important in locating subsurface reef trends and delineating exploration targets in the reef and dolomitized back-reef, shelf-margin deposits.

Geological Setting and Carbonate Facies

Throughout Permian times the Delaware Basin was a structural low and a site of basinal deposition, while the Central Basin Platform and the Northwestern Shelf were positive areas of shallow shelf deposition (Harms, 1974). Carbonate reefs and banks developed along the margin of the Delaware Basin, and these are now partially exposed in the Guadalupe Mountains. A general stratigraphic column for the Permian shelf-to-basin sequence is given in Figure 3-70.

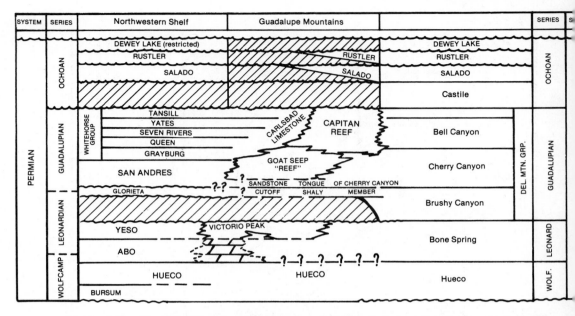

Figure 3-70 Stratigraphic column and rock-unit correlation chart for Permian reef, shelf, and basin deposits of southeast New Mexico (after LeMay, 1972).

The rock units shown in the stratigraphic column can be subdivided into a number of facies characteristic of deposition across a shelf-to-basin profile. These facies are summarized in Wilson (1975, p. 222–223) after Tyrrell (1969) and Dunham (1972).

The shelf or platform facies are represented by the rock formations of the Northwestern Shelf, including the San Andres, Grayburg, Queen, Seven Rivers, Yates, Tansill, and Carlsbad (Fig. 3-70). Across the shelf toward the basin margin the following facies are encountered:

1 Red shale and siltstone.
2 Evaporites, including halite, gypsum, and anhydrite, fine sandstones, and dolomitic wackestones.
3 Dolomitic wackestones with ostracodes and calcispheres, laminated mudstones interpreted as algal stromatolites, peloidal and dolomitic grainstones.
4 Skeletal lithoclast dolomite grainstones with coated grains, foraminifera, dasycladacean algae, oöids, and gastropods (see Fig. 3-71 for faunal distributions).

Many studies of the deposits across the shelf suggest extensive alteration by early diagenesis, including alterations associated with periods of subaerial exposure (Dunham, 1969a, 1969b). Deposition of shelf sediments is also interpreted as being cyclical (Silver and Todd, 1969; Meissner, 1972); however, there is still much disagreement in interpreting the shelf sequences and their relationship to sea level throughout the Permian (Wilson, 1975, p. 230–231; Esteban and Pray, 1976).

The basin-margin facies of the Delaware Basin include the so-called organic reef buildups represented by the Capitan Limestone (massive phase) and earlier deposits such as the Goat Seep, Getaway "Bank," Victorian Peak, and Abo reef trends. The reef

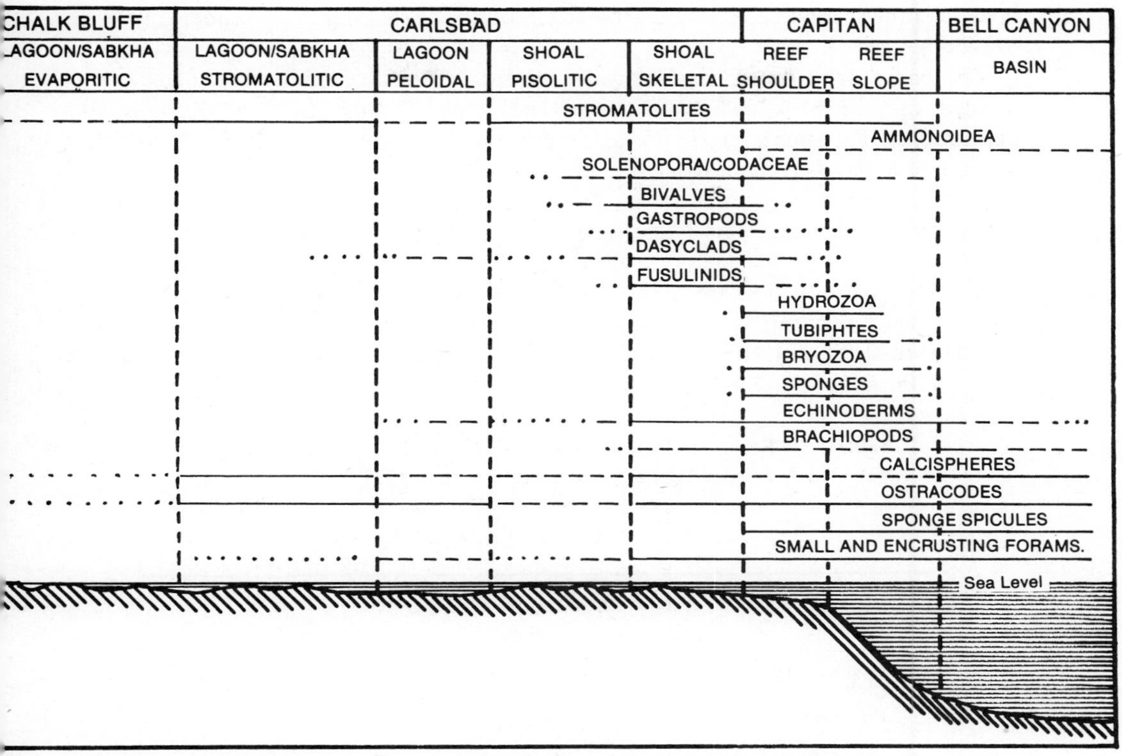

Example 11   193

CHALK BLUFF	CARLSBAD				CAPITAN		BELL CANYON
LAGOON/SABKHA EVAPORITIC	LAGOON/SABKHA STROMATOLITIC	LAGOON PELOIDAL	SHOAL PISOLITIC	SHOAL SKELETAL	REEF SHOULDER	REEF SLOPE	BASIN

STROMATOLITES
AMMONOIDEA
SOLENOPORA/CODACEAE
BIVALVES
GASTROPODS
DASYCLADS
FUSULINIDS
HYDROZOA
TUBIPHTES
BRYOZOA
SPONGES
ECHINODERMS
BRACHIOPODS
CALCISPHERES
OSTRACODES
SPONGE SPICULES
SMALL AND ENCRUSTING FORAMS.
Sea Level

Figure 3-71 Distribution of common fossil groups across the shelf to basin profile of the Permian reef complex, Texas and New Mexico (after Schmidt, 1977).

trends show a basinward migration with time. The organic reef facies consists of massive skeletal lithoclast grainstones and wackestones, bioclastic debris mixed with lime mud or silt and boundstones composed of sessile benthos and encrusting biota. Fauna includes bryozoans, red algae, *Tubiphytes*, foraminifera, sponges, brachiopods, and crinoids (Fig. 3-71). Associated diagenetic alterations include void-filling cements, collapsed beds, and veins filled by calcite spar. These features are interpreted as the result of marine or meteoric vadose diagenesis (Dunham, 1969a). The reef limestones are locally recrystallized and dolomitized so that details of their original composition are obliterated.

Disagreement still exists as to whether the carbonate buildups around the rim of the Delaware Basin are true organic reefs or quiet water mudmounds (Newell et al., 1953; Dunham, 1972). The destruction of early depositional features by diagenesis is intense, and the lack of obvious organic framework together with the abundance of micrite in the "reefal" limestones has led to the alternative mudmound hypothesis.

The talus or fore-slope facies which is adjacent to the "reef" is represented by Capitan Limestone with steep dipping beds. The fore-slope rocks include basinward-extending tongues of breccia with worn and broken fossils and boulders derived from the shelf-rim limestones. These coarse deposits are associated with partially dolomitized fine lithoclast skeletal wackestones (Tyrrell, 1969; Dunham, 1972; Wilson, 1975, p. 223). At the toe of the slope dark, well-bedded lithoclast skeletal wackestone-packstones of the Delaware Mountain Group occur associated with slumps, boulder beds, channels, load casts, and carbonate mounds (Friedman and Sanders, 1978, p. 378–380).

The basin deposits of the Delaware Mountain Group include fine, black, well-bedded wackestone-packstones with sponge spicules, foraminifera, radiolarians, and ammonoids. The starved basin phase is represented by radioactive silty shales and thin carbonate beds (Tyrrell, 1969; Dunham, 1972).

Reservoirs in Permian Reef Complex

Hydrocarbon production from the Permian reef complex is mostly from the shelf dolomites and reef limestones, with additional but less important production from the coarse, fore-slope talus material and channel sandstones within the basin. Three fields from the Permian sequence are summarized in Tables 3-7, 3-8, and 3-9. These fields—

Table 3-7   Reservoir Data for the Empire Abo Field, Eddy County, New Mexico (from Sigma Seismic Surveys Inc., 1977)

	EMPIRE ABO FIELD
Hydrocarbon Type:	Oil and associated gas
Reservoir Type:	Biohermal (reef) mass
Formation:	Abo Formation
Age:	Permian (Leonardian)
Lithology:	Light tan, gray-gray white, fine to coarse crystalline, sucrosic, slight to very anhydritic, clean dolomite
Geologic Setting: (Original)	Barrier reef of Permian age along northern hinge line of Delaware Basin, which has been dolomitized and recrystallized
Reservoir Characteristics:	Porosity is intercrystalline with vugular porosity common, porosity and permeability are lowest in western part of field due to higher percentage of anhydrite
Trapping Mechanism:	Permeability barrier on north Dolomite becomes more dense updip to the west Oil-water contact at south end and east end of field
Pay Thickness:	15-726 feet
Porosity:	1.5-18.3%
Permeability:	0.1-1,970 md
Pay Zone Depth:	5,000-6,000 feet
Oil Gravity:	43° gravity API corrected sweet crude
Gas Data:	0.889° gravity semi-sour to sour. Increasing H_2S content with depth

Example 11   **195**

Table 3-8   Reservoir Data for the Kemnitz Field, Lea County, New Mexico (from Sigma Seismic Surveys Inc., 1977)

	KEMNITZ FIELD
Hydrocarbon Type:	Oil
Reservoir Type:	Barrier reef
Formation:	Lower Wolfcamp
Age:	Permian
Lithology:	Reefal limestone
Geologic setting: (Original)	An accretionary carbonate mass which developed on a shelf margin separating the Delaware Basin from the Northwest Shelf
Reservoir Characteristics:	The reef-wall facies porosity is largely unaffected by spar-cementation. Some dolomite rhombs are present in large pores and vugs. The backreef and forereef facies are generally porous and permeable close to the reef.
Trapping Mechanism:	Barrier-reef development with updip loss of porosity within backreef units
Pay Thickness:	80-100 feet
Porosity:	18% maximum; median value 8.5%
Permeability:	Less than 1.0 md to more than 1,000 md, median value 5 md
Pay Zone Depth:	10,500-11,000 feet
Oil Gravity:	39° API

the Empire Abo, Kemnitz, and Cato San Andres—provide examples of production from both reef trends and shelf dolomites.

The Empire Abo field produces from light tan to grey dolomite and anhydritic dolomite of the Abo Formation, which contains few recognizable fossils. The reservoir rock has good secondary porosity as the result of a well-developed fracture system together with associated vuggy porosity and intercrystalline porosity between dolomite rhombs. The reservoir facies is interpreted as part of a reef trend (Fig. 3-72), separating the shales and finely crystalline anhydritic dolomites of the shelf facies from the black and brown argillaceous, cherty carbonates, and interbedded fine sandstones interpreted as basin deposits (LeMay, 1972). Once a reefal model was applied to the Empire Abo Field, a successful reef trend exploration program resulted in a number of additional discoveries along the Abo reef trend (Fig. 3-72).

The Kemnitz field is developed over a north-south trending ridge at the margin of the Delaware Basin. The reservoir is a stratigraphic trap in an early Wolfcamp shelf-margin reefal limestone which loses porosity updip. The limestone buildup is

Table 3-9   Reservoir Data for the Cato San Andres Field,
Chaves County, New Mexico (from Sigma Seismic Surveys Inc., 1977)

	CATO SAN ANDRES FIELD
Hydrocarbon Type:	Oil
Reservoir Type:	Porosity development in dolomite
Formation:	San Andres
Age:	Permian (Guadalupian)
Lithology:	Buff to brown anhydritic dolomite
Geologic Setting: (Present)	Gentle southeast dip with apparent reversal on extreme southeast portion of the field
Reservoir Characteristics:	Dolomite contains intercrystal and vuggy porosity with vertical and horizontal fractures
Trapping Mechanism:	Gradual permeability loss updip to tighter more anhydritic dolomite
Pay Thickness:	Gross average 111 feet: net average 33 feet
Porosity:	7.5-8.5%
Permeability:	0.1-1.5 md
Pay Zone Depth:	3,300-3,600 feet
Oil Gravity:	25°-27° API: sour

asymmetrical and thins gradually shoreward where it passes into back-reef grainstone-wackestones composed of green algal fragments, skeletal grainstones and packstones, and dasyclad packstones and wackestones which contain peloids, foraminifera, and dasycladacean algae in a micritic matrix. These back-reef facies have some porosity between particles where micrite and calcite spar does not occur (Malek-Aslani, 1970).

The reef facies forms the main reservoir horizon and is composed of a framework of *Tubiphyte* remains associated with fusulinids, crinoid ossicles, bryozoans, and brachiopods. Micrite has partially filled interparticle porosity but the rock has remained relatively porous and permeable (Table 3-8). The reef buildup terminates abruptly at the shelf margin on the edge of the Delaware Basin (Malek-Aslani, 1970). The fore-reef deposits are a mixture of material derived from the shelf and dark-colored, fine-grained basin deposits. As with the Abo trend, a number of important oil fields have been found along the barrier-reef trend associated with the Kemnitz field.

The Cato San Andres field, in contrast to the other two examples, occurs in shelf facies associated with shelf anhydrites. The field produces from anhydritic dolomites (Table 3-9). The Cato San Andres is a classic stratigraphic trap, and the structural contours on top of the San Andres Formation show no evidence of closure above the field (Fig. 3-73). The dolomitic reservoir rock is sandwiched between impervious

Example 11   197

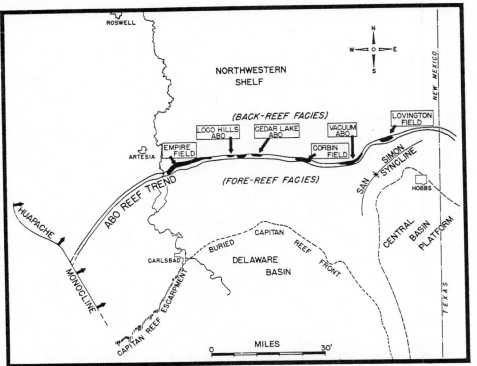

Figure 3-72   Location map of the Abo reef trend and related oil fields (after LeMay, 1972).

anhydrite and loses porosity updip where it passes into more anhydritic dolomite and anhydrite (Lanan, 1981).

The dolomitic reservoir rock is light brown in color and contains evidence of algal-mat structures, stromatolites, intraclasts, mud cracks, and bird's-eye structures (Lanan, 1981). Most of the porosity is intercrystalline between dolomite rhombs but moldic porosity and minor fracture porosity occur. The moldic porosity is related to the leaching of carbonate skeletal fragments. Anhydrite is the main void-filling mineral and the distribution of porous reservoir rock is closely linked to lack of anhydrite.

Conclusions

The Permian reef complex of Texas and New Mexico has been cited as a classic example of a barrier-reef sequence that passes from restricted shelf or lagoonal deposits composed of clastics, evaporites, dolomites, and limestones into a reefal facies at the margin of the shelf, and finally into restricted basin deposits. Reservoirs in the sequence are found in the shelf dolomites, within the reef trends, in the coarse fore-slope deposits, and even in deep basinal sandstones.

Although this sequence has been extensively studied, both in the subsurface and in outcrop, the interpretation of many important features such as the nature of the "reefal" buildups and the extent of subaerial exposure and early freshwater diagenesis of the shelf, reef, and slope deposits is still in dispute.

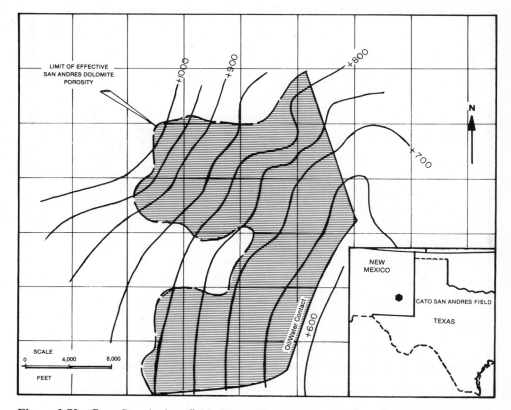

Figure 3-73   Cato San Andres field, Chave County, New Mexico, showing structure contours on top of the San Andres Formation, contour interval is 50 feet (after Sigma Seismic Surveys Inc., 1977).

EXAMPLE 12.   A CARBONATE TURBIDITE RESERVOIR—THE "SCAGLIA CALCAIRE" (CRETACEOUS-TERTIARY) OF CENTRAL ITALY

B. Duvernoy, and J. Reulet

The Scaglia Calcaire, a Cenomanian to Lower Eocene deposit (Fig. 3-74), was studied in field sections from the Marches Chain and in offshore wells in the Adriatic Sea. In the offshore the formation was penetrated below the Tertiary cover of the Marches Basin (Fig. 3-75).

Field sections showed the Scaglia Calcaire to be a deep marine deposit, and in the upper part to contain brecciola horizons of Campanian, Maestrichian, and Paleocene age. The term brecciola is defined as a rubble of carbonate rock, usually angular, interstratified with dark-colored marine shales (Friedman and Sanders, 1978). These brecciolas are interpreted as the deposits of turbidity currents.

The boreholes were used to study variations in petrophysical characters and diagenetic evolution of the reservoir horizons within the brecciolas of the Scaglia Calcaire.

Example 12 199

Sedimentary Facies

The Scaglia Calcaire is composed of alternating argillaceous limestones with pelagic fauna, and brecciola beds with benthonic fauna.

The argillaceous limestones consist of pelagic foraminiferal and coccolith wacke-stones. The rocks are very fine grained, and individual beds are 5 to 20 cm thick, separated by clay films. Continuous beds of chert or nodules containing ghosts of pelagic foraminifera are common. The argillaceous limestones vary in color from white to brick-red and contain common cemented and uncemented microfissures.

The brecciola layers are interbedded with the argillaceous limestones. The brecciola beds have varying thicknesses, which range from a few centimeters to several meters. Their base shows distinct graded bedding and has an abrupt contact with the underlying argillaceous limestones. In places this contact may be erosional into the underlying limestones. Sedimentary structures such as flute casts and groove casts are common at this basal contact. The contact between the brecciolas and the overlying limestone is gradational, and the sequence as a whole has a rhythmic nature.

The clasts that form the brecciolas are most commonly packstones with benthonic fauna, mainly rudistids, orbitoidids, echinoderms, encrusting red algae, and rare fragments of corals and bryozoans. This fauna all originated from a neritic carbonate shelf and was contemporaneous with the deep basin sediments in which the brecciolas are interbedded. All clasts are of carbonate and the individual beds grade upward from coarse calcirudite to fine calcarenite.

Lateral and Vertical Evolution

The large-scale sequence that passes from marls with trace fossils (Aptian to Albian) to the top of the Scaglia (Paleocene to Early Eocene) is shown in Figure 3-76. It is a coarsening-upward or negative sequence related to the infilling with time of the depositional basin. The small-scale sequences associated with each brecciola horizon are fining-upward sequences composed of the following:

> TOP
> Lime mudstone composed of coccoliths, coccolith debris, and small foraminifera.
> Lime wackestone with pelagic foraminifera.
> Brecciola with graded bedding and benthonic fauna.
> BASE

The upper two lithologies do not necessarily occur superimposed, but may alternate or occur as lateral facies. The sequence as a whole has a varied thickness from less than 1 m to more than 10 m.

Figure 3-77 illustrates the depositional setting of the brecciola sequences. The brecciolas are best developed in the south and southeast and thin to the north and northwest, where they become less common and finally disappear. The brecciolas form a continuous sheet to the south-southeast, which passes into discontinuous channels and tongues in the central part of the basin. The brecciolas also change in thickness from the south-southeast, where the thickest sequence with the coarsest basal beds is

Stages	Biostratigraphic Units	Lithostratigraphic Units
QUATERNARY	Ammonia beccarii / Elphidium spp. — Globigerina pachyderma / Cassidulina laevigata carinata	
CALABRIAN	Hyalinea balthica	
UPPER PLIOCENE	Globorotalia inflata / Anomalina ornata	
MIDDLE PLIOCENE	Globorotalia gr. crotonensis (présence d'olistostromes)	
LOWER PLIOCENE	Globorotalia puncticulata / Globorotalia hirsuta	
UPPER MIOCENE	Oligotypique petits Coprolithes de crustacés	SCHLIER
TORTONIAN	Globorotalia menardii	MARNOSO ARENACEA (Ombrie)
SERRAVILLIAN LANGHIAN	Globorotalia gr. mayeri - acrostoma / Globoquadrina spp.	BISCIARIO
AQUITANIAN	Catapsydrax dissimilis - Globigerinoides trilobus	SCAGLIA CINEREA
OLIGOCENE	Cibicides cushmani - Planulina renzi	

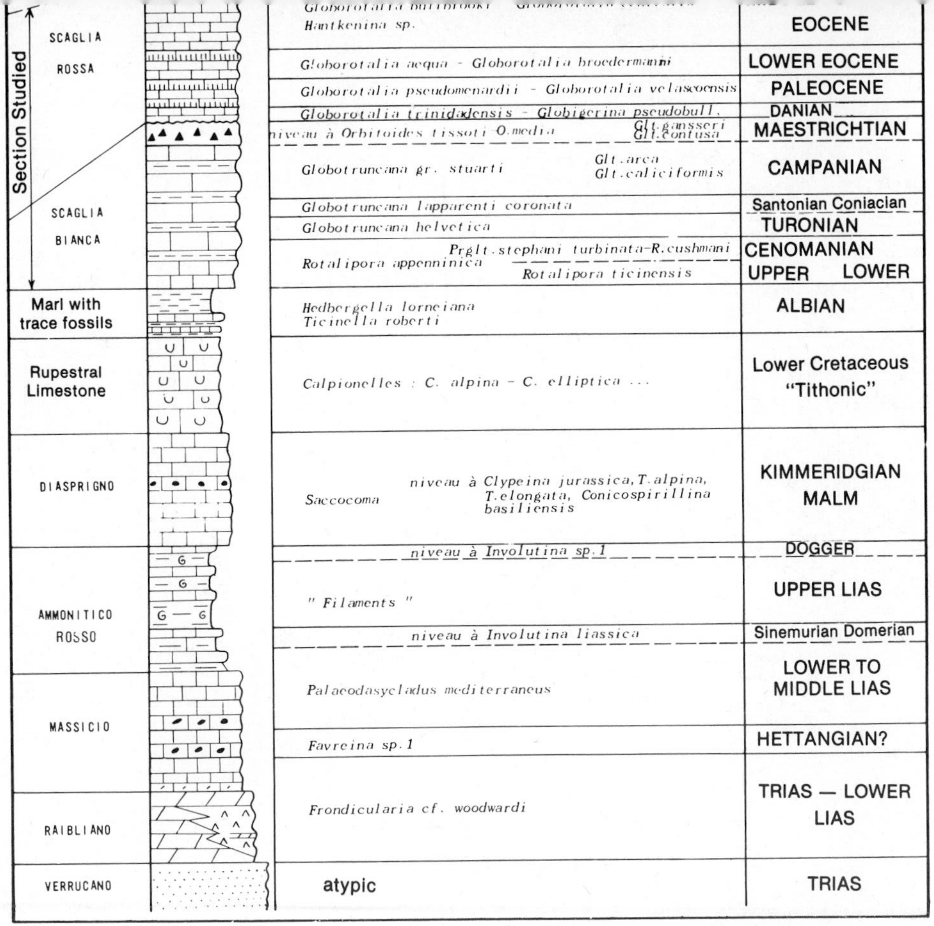

Lithostratigraphic unit	Lithology	Biostratigraphy (fossils)	Age
SCAGLIA ROSSA		*Globotruncana ...* *Hantkenina sp.*	EOCENE
		Globorotalia aequa - Globorotalia broedermani	LOWER EOCENE
		Globorotalia pseudomenardii - Globorotalia velascoensis	PALEOCENE
		Globorotalia trinidadensis - Globigerina pseudobull...	DANIAN
		niveau à *Orbitoides tissoti O.media* / *Glt. gansseri* *Glt.gigantusa*	MAESTRICHTIAN
SCAGLIA BIANCA		*Globotruncana gr. stuarti* / *Glt. arca* *Glt.calciciformis*	CAMPANIAN
		Globotruncana lapparenti coronata	Santonian Coniacian
		Globotruncana helvetica	TURONIAN
		Praelt.stephani turbinata-R.cushmani / *Rotalipora appenninica* / *Rotalipora ticinensis*	CENOMANIAN UPPER LOWER
Marl with trace fossils		*Hedbergella lorneiana* / *Ticinella roberti*	ALBIAN
Rupestral Limestone		*Calpionelles : C. alpina - C. elliptica ...*	Lower Cretaceous "Tithonic"
DIASPRIGNO		niveau à *Clypeina jurassica,T.alpina, T.elongata, Conicospirillina basiliensis* / *Saccocoma*	KIMMERIDGIAN MALM
		niveau à *Involutina sp.1*	DOGGER
AMMONITICO ROSSO		"Filaments"	UPPER LIAS
		niveau à *Involutina liassica*	Sinemurian Domerian
MASSICIO		*Palaeodasycladus mediterraneus*	LOWER TO MIDDLE LIAS
		Favreina sp.1	HETTANGIAN?
RAIBLIANO		*Frondicularia cf. woodwardi*	TRIAS — LOWER LIAS
VERRUCANO		atypic	TRIAS

Figure 3-74 Principal lithostratigraphic and biostratigraphic units of the Scaglia Calcaire.

201

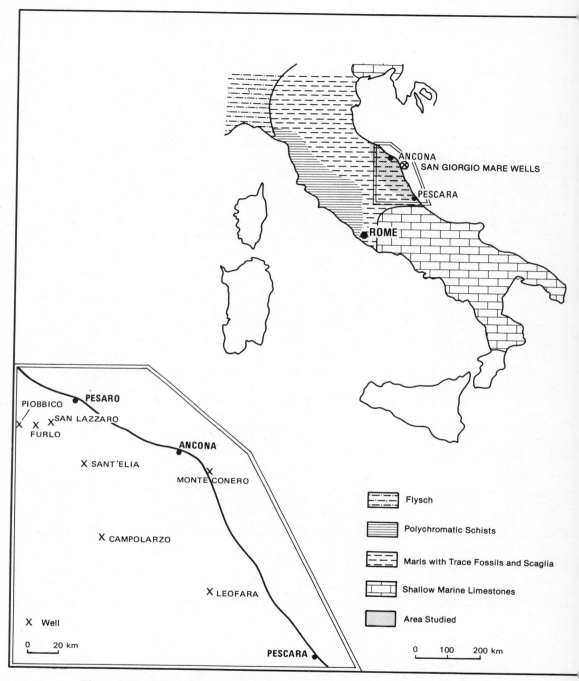

Figure 3-75 Upper Cretaceous facies distribution and location map, central Italy.

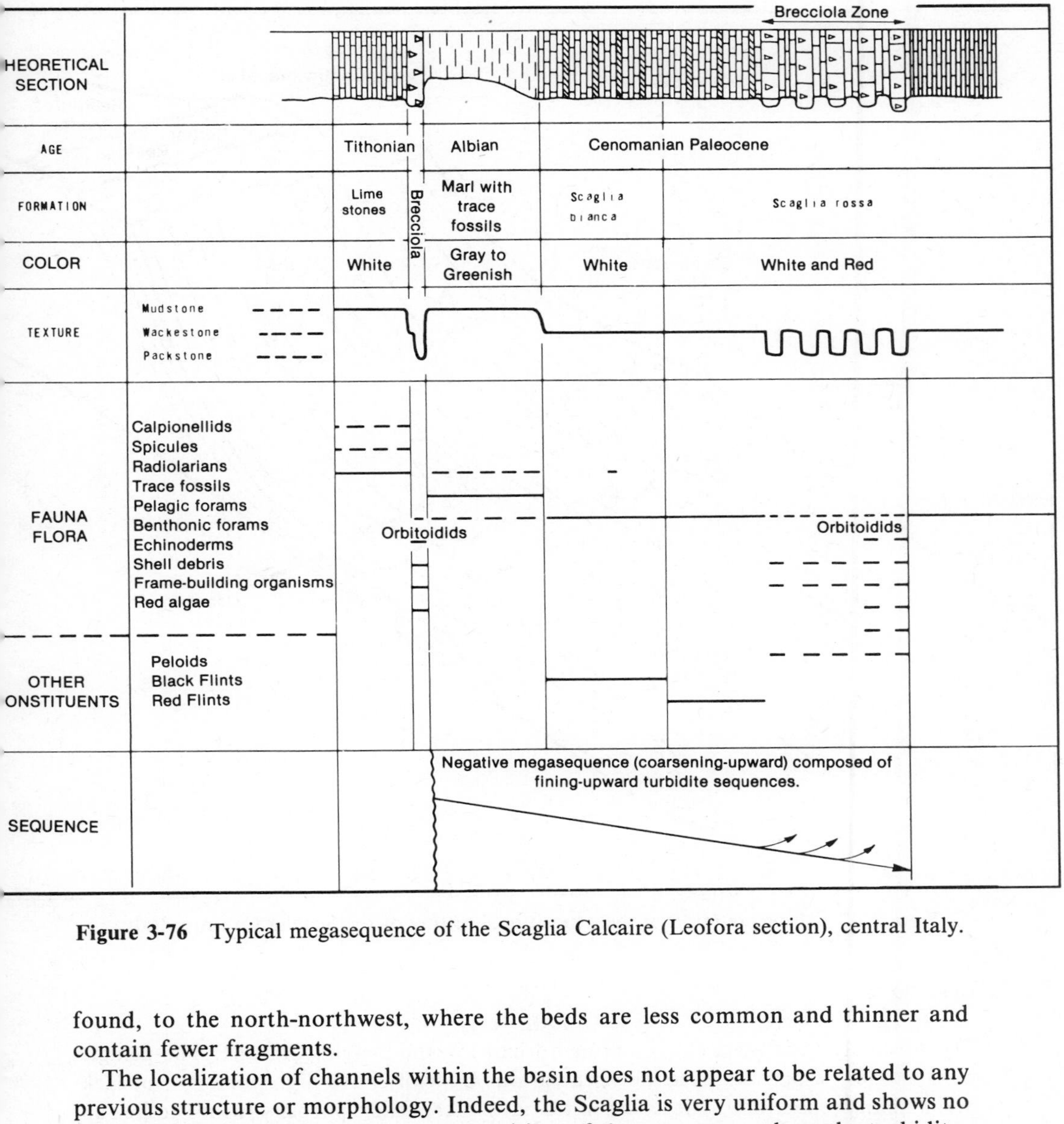

Example 12 203

Brecciola Zone

THEORETICAL SECTION					
AGE		Tithonian	Albian	Cenomanian	Paleocene
FORMATION		Lime stones	Brecciola / Marl with trace fossils	Scaglia bianca	Scaglia rossa
COLOR		White	Gray to Greenish	White	White and Red
TEXTURE	Mudstone / Wackestone / Packstone				
FAUNA FLORA	Calpionellids / Spicules / Radiolarians / Trace fossils / Pelagic forams / Benthonic forams / Echinoderms / Shell debris / Frame-building organisms / Red algae	Orbitoidids			Orbitoidids
OTHER CONSTITUENTS	Peloids / Black Flints / Red Flints				
SEQUENCE		Negative megasequence (coarsening-upward) composed of fining-upward turbidite sequences.			

Figure 3-76 Typical megasequence of the Scaglia Calcaire (Leofora section), central Italy.

found, to the north-northwest, where the beds are less common and thinner and contain fewer fragments.

The localization of channels within the basin does not appear to be related to any previous structure or morphology. Indeed, the Scaglia is very uniform and shows no variations that would permit the recognition of deeper zones, where the turbidites would have been channeled. The location of the channels may have depended more on the morphology of the shelf.

A major period of tectonism during the Middle Pliocene has created long, approximately north-south anticlinal trends. These anticlines are asymmetrical and are overturned and thrust along fault planes on the eastern limb (Fig. 3-78). The axis of these anticlines intersects the brecciola channel fills at a small angle. The brecciolas in the wells studied were more localized on the western limbs and on the tops of the structures. This configuration is important in creating the reservoirs.

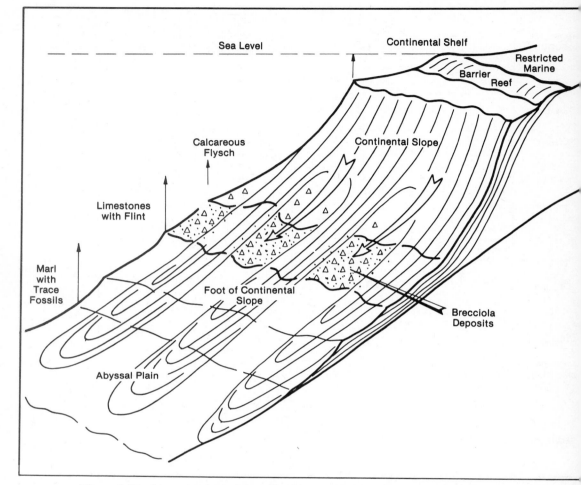

Figure 3-77   Interpretation of the depositional setting of the Scaglia Calcaire, Italy.

Reservoirs

The brecciola beds form the reservoir horizons in the Scaglia Calcaire. The pelagic wackestones, which are extremely fine grained, locally have high porosity but lack permeability (chalklike). The calcarenites have interparticle porosity from 1 to 2%, up to 22% in the clean calcarenites. The permeability in these is high.

   Dissolution and/or cementation occurred during diagenesis and was controlled by the original porosity and permeability. The secondary porosity is linked to the dissolution of tests of organisms, breccia clasts, or micrite. Penecontemporaneously, rim cement and equigranular calcite mosaics filled pore spaces. Cementation occurred in patches or developed throughout the rock, totally destroying the primary porosity. Cementation reduced porosity, particularly in fine calcarenites.

   Tectonism has had a considerable influence on reservoir formation. Fractures have been opened that are conducive to the flow of undersaturated waters. The zones of stress on the western sides of the structures are sites of open fractures, whereas in zones

Example 12 205

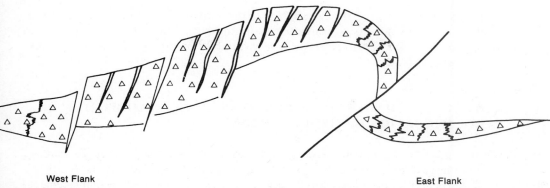

West Flank

ension — open fractures

Vater circulation and dissolution

igh primary porosity (coarse particles)

arge reservoir thickness

East Flank

Compression — stylolites, closed fractures

Poor water circulation, cementation

Low primary porosity (fine grained)

Thin reservoir horizon

Figure 3-78 Position of reservoir in relation to structure, Scaglia Calcaire, Italy.

of compression on the eastern side of the folds stylolites developed and cements were precipitated. The fractures and faults influence the drainage of the reservoirs throughout the entire formation. They link the different calcarenitic layers to form a single rather than a multibedded reservoir. The differences shown in wells drilled on the same structure can be attributed to the presence of limestone brecciolas with primary porosity. The asymmetrical structures produce reservoir anisotropy because of the distribution of open versus closed fractures and the changing thickness of the brecciola horizon across the structure (Fig. 3-78).

Conclusions

The Scaglia reservoir is formed by limestone brecciolas or calcarenites. These occur as graded beds composed of debris brought in from the shallow shelf. The debris is deposited in a deep-water environment interbedded with fine sediments containing pelagic fauna.

The brecciolas were emplaced by turbidity currents that transported the debris. This transport resulted in disruption of shelf deposition and the presence of breaks within the talus bordering the shelf. As a result of the respective position of the shelf and the basin, the turbidity currents moved in a south-north direction in the middle of the basin and in a southwest-northeast direction elsewhere. The reservoirs are characterized by:

1 A primary interparticle porosity.
2 A diagenetic porosity caused by solution.
3 Fracture porosity.

The fracture porosity is of tectonic origin. Cementation is contemporaneous with dissolution and can locally obliterate porosity. The cementation is more likely to occur in compressional zones on the eastern side of the anticlinal structures where pressure solution is dominant. As a consequence of the structural setting, hydrocarbon production from this kind of reservoir is very variable, ranging from virtually nothing to very large volumes.

EXAMPLE 13.   DIAGENETIC RESERVOIRS ASSOCIATED WITH EARLY MIGRATION OF HYDROCARBONS

When migration of petroleum is early and predates an episode of tilting or folding there is a need to be aware of the possibility of diagenetic traps (Wilson, H. H., 1975, 1977). Early migration is indicated by many carbonate sequences where emplacement of oil has obviously preceded diagenetic alterations, especially cementation.

Figure 3-79 depicts the sequence of events associated with the formation of a diagenetic trap. In the example shown the original trapping mechanism is loss of porosity updip as the result of a facies change from oölitic grainstone to lime mudstone. At the oil/water contact in reservoirs conditions are often favorable for extensive precipitation of calcite cements. This is a common phenomenon in many carbonate oil reservoirs, which may show no evidence of cementation except in the region of the oil/water contact. The occurrence of calcite cement in this position has

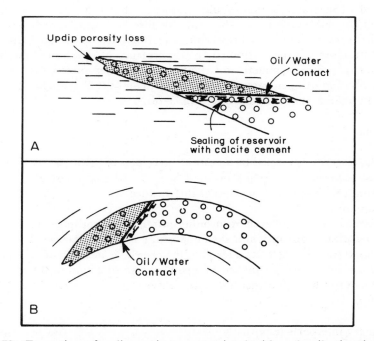

Figure 3-79   Formation of a diagenetic trap associated with early oil migration. (*A*) Accumulation of hydrocarbons in a stratigraphic trap as the result of porosity loss updip. Heavy cementation by calcite at the oil/water contact seals in the oil. (*B*) With folding of the strata the oil/water contact remains fixed so that the final position of the reservoir is unrelated to structure.

been related to the action of sulfate-reducing bacteria and explained using the following equation (Friedman and Sanders, 1978, p. 130, 131, 158):

$$\text{Ca}^{2+} + \text{SO}_4^{2-} + 2\ \text{CH}_2\text{O} \xrightarrow{\textit{Bacteria}} \text{CaCO}_3 + \text{H}_2\text{O} + \text{H}_2\text{S}$$

Formation   Organic   Calcite
waters   matter   cement

The calcite cement will eventually fill all the pore spaces at the oil/water contact, sealing the oil within the reservoir and making the reservoir impermeable to further fluid movement.

When extensive cementation at the oil/water contact is followed by a period of folding or tilting there will be no loss of oil from the reservoir and no relocation of oil to structural highs (Fig. 3-79). When a diagenetic trap of this kind is present within a sequence of rocks, an exploration program of drilling on structural highs will be unsuccessful. To locate potential diagenetic traps, it is necessary to untilt or unfold the sequence, using information from seismic sections, so as to find the likely position of early structural or stratigraphic traps, which may then have been sealed by cementation at the oil/water contact.

REFERENCES

Babcock, J. A., 1977, Calcareous algae, organic boundstones, and the genesis of the upper Capitan Limestone (Permian, Guadalupian), Guadalupe Mountains, west Texas and New Mexico, p. 3–44, in E. M. Hilman, and S. J. Mazullo, Eds., *Upper Guadalupian Facies, Permian Reef Complex, Guadalupe Mountains, New Mexico and Texas:* Midland Texas, Society of Economic Paleontologists and Mineralogists, Permian Basin Section, Pub. 77–16, 508 p.

Bebout, D. G., and Schatzinger, R. A., 1978, Distribution and geometry of an oölite-shoal complex—Lower Cretaceous Sligo Formation, south Texas: *Trans. Gulf Coast Assoc. Geol. Soc.* v. 28, p. 33–45.

Becher, J. W., and Moore, C. H., 1976, The Walker Creek Field: a Smackover diagenetic trap: *Trans. Gulf Coast Assoc. Geol. Soc.,* v. 26, p. 34–56.

Bentley, B. P., 1979a, Carbonate lithofacies and diagenetic features of the Guelph Formation (Middle Silurian) in the Amoco Production Berg-Berg, 1-21 unit well, Presque Isle County, Michigan: unpublished M.S. thesis, Rensselaer Polytechnic Institute, 147 p.

Bentley, B. P., 1979b, Carbonate lithofacies and diagenetic features of the Guelph Formation (Middle Silurian) in the Amoco Production Berg-Berg 1-21 unit well, Presque Isle County, Michigan: *The Compass of Sigma Gamma Epsilon,* v. 57, p. 16–26.

Bishop, W. F., 1968, Petrology of the upper Smackover Limestone in the North Haynesville field, Claibourne Parish, Louisiana: *Bull. Am. Assoc. Pet. Geol.* v. 52, p. 92–128.

Briggs, L. I., and Briggs, D., 1974, Niagara-Salina relationships in the Michigan Basin, p. 1–23, in R. V. Kesling, Ed., *Silurian Reef-Evaporite Relationships:* Michigan Basin Geological Society Field Conference, 111 p.

Brigham, R. J., 1971, Structural geology of southwestern Ontario and southeastern Michigan: *Ontario Dept. Mines and Northern Affairs, Petroleum Resources Sec.,* Paper 71-2, 110 p.

Bushaw, D. J., 1968, Environmental synthesis of the east Texas Lower Cretaceous: *Trans. Gulf Coast Assoc. Geol. Soc.* v. 18, p. 416–438.

Cohee, G. V., and Landes, K. K., 1958, Oil in the Michigan Basin, p. 473–493, in L. G. Weeks, Ed., *Habitat of Oil:* Tulsa, Okla., American Association of Petroleum Geologists, 1384 p.

Collins, S. E., 1980, Jurassic Cotton Valley and Smackover reservoir trends, east Texas, north Louisiana, and south Arkansas: *Bull. Am. Assoc. Pet. Geol.,* v. 64, p. 1004–1013.

Cook, T. D., 1979, Exploration history of south Texas Lower Cretaceous carbonate platform: *Bull. Am. Assoc. Pet. Geol.* v. 63, p. 32–49.

Cussey, R., and Friedman, G. M., 1977, Patterns of porosity and cement in oöid reservoir in Dogger (Middle Jurassic) of France: *Bull. Am. Assoc. Pet. Geol.,* v. 61, p. 511–518.

Dickinson, K. A., 1968, Upper Jurassic stratigraphy of some adjacent parts of Texas, Louisiana, and Arkansas: *U.S. Geol. Survey, Prof. Paper 594E,* 25 p.

Dickinson, K. A., 1969, Upper Jurassic carbonate rocks in northeastern Texas and adjoining parts of Arkansas and Louisiana: *Trans. Gulf Coast Assoc. Geol. Soc.,* v. 19, p. 175–186.

Dunham, R. J., 1962, Classification of carbonate rocks according to depositional texture, p. 108–121, in W. E. Ham, Ed., *Classification of Carbonate Rocks:* Tulsa, Okla., American Association of Petroleum Geologists, Mem., 1, 279 p.

Dunham, R. J., 1969a, Early vadose silt in Townsend Mound (reef), New Mexico, p. 139–181, in G. M. Friedman, Ed., *Depositional environments in carbonate rocks, a symposium:* Tulsa, Okla. Society of Economic Paleontologists and Mineralogists, Spec. Pub. No. 14, 209 p.

Dunham, R. J., 1969b, Vadose pisolites in the Capitan reef (Permian), New Mexico and Texas, p. 182–191, in G. M. Friedman, Ed., *Depositional environments in carbonate rocks, a symposium:* Tulsa, Okla., Society of Economic Paleontologists and Mineralogists, Spec. Pub. No. 14, 209 p.

Dunham, R. J., 1972, Capitan reef, New Mexico and Texas: facts and questions to aid interpretation and group discussion: Midland, Texas, Society of Econ. Paleontologists and Mineralogists, Permian Basin Section, Pub. 72–14, 270 p.

Embry, A. F., and J. E., Klovan, J. E., 1971, A Late Devonian reef tract on northeastern Banks Island, N.W.T.: *Bull. Can. Pet. Geol.,* v. 19, p. 730–781.

Esteban, M., and Pray, L. C., 1976, Nonvadose origin of pisolitic facies, Capitan reef complex (Permian), Guadalupe Mountains, New Mexico and west Texas: *Bull. Am. Assoc. Pet. Geol.,* v. 60, p. 670.

Friedman, G. M., and Sanders, J. E., 1978, *Principles of Sedimentology:* New York, John Wiley and Sons, 792 p.

Fuller, J. G. C. M., 1961, Ordovician and contiguous formations in North Dakota, South Dakota, Montana and adjoining areas of Canada and the United States: *Bull. Am. Assoc. Pet. Geol.,* v. 45, p. 1334–1363.

Gill, D., 1977, Salina A-1 sabkha cycles and the Late Silurian paleogeography of the Michigan Basin: *J. Sediment. Pet.,* v. 47, p. 979–1017.

Gill, D., 1979, Differential entrapment of oil and gas in Niagaran pinnacle-reef belt of northern Michigan: *Bull. Am. Assoc. Pet. Geol.,* v. 63, p. 608–620.

Harms, J. C., 1974, Brushy Canyon Formation, Texas: a deep-water density current deposit: *Bull. Geol. Soc. Am.,* v. 85, p. 1763–1784.

Herrmann, L. A., 1971, Lower Cretaceous Sligo reef trends in central Louisiana: *Trans. Gulf Coast Assoc. Geol. Soc.,* v. 21, p. 187–198.

Huh, J. M., Briggs, L. I., and Gill, D., 1977, Depositional environments of pinnacle reefs, Niagaran and Salina Groups, northern shelf, Michigan Basin, p. 1–21, in J. H. Fisher, Ed., *Reefs and evaporites— concepts and depositional models:* Tulsa, Okla., American Association of Petroleum Geologists, Studies in Geology, No. 5, 196 p.

King, P. B., 1948, Geology of the southern Guadalupe Mountains, Texas: *U. S. Geol. Survey, Prof. Paper 215,* 183 p.

Klüpfel, W., 1917, Üeber die Sedimente der Flachsee in Lothringer Jura: *Geol. Rundsch.,* v. 7, p. 97–109.

Lanan, H. K., 1981, *Depositional facies, porosity, and permeability in Cato field, Chaves County, New Mexico:* unpublished M.A. thesis, University of Texas, 275 p.

LeMay, W. J., 1972, Empire Abo field, southeast New Mexico, p. 472–480, in R. E. King, Ed., *Stratigraphic oil and gas fields—classification, exploration methods, and case histories:* Tulsa, Okla., American Association of Petroleum Geologists, Mem. 16, 687 p.

Malek-Aslani, M., 1970, Lower Wolfcampian reef in Kemnitz field, Lea County, New Mexico: *Bull. Am. Assoc. Pet. Geol.,* v. 54, p. 2317–2335.

Mantek, W., 1973, Niagaran pinnacle reefs in Michigan: Michigan Basin Geological Society Annual Field Conference, Guidebook, p. 35–46.

Mathis, R. L., 1978, *Carbonate sedimentation and diagenesis of reef and associated shoal-water facies, Sligo Formation (Aptian), Black Lake field, Natchitoches Parish, Louisiana:* Unpublished M.S. thesis, Rensselaer Polytechnic Institute, 214 p.

McFarlan, E., Jr., 1977, Lower Cretaceous sedimentary facies and sea level changes, U.S. Gulf Coast, p. 5–11 in D. G. Bebout and R. G. Loucks, Eds., *Cretaceous carbonates of Texas and Mexico:* University of Texas, Austin, Bureau of Economic Geology, Rept., Inv. 89, 332 p.

Meissner, F. F., 1972, Cyclic sedimentation in Middle Permian strata of the Permian Basin, west Texas and New Mexico, in *Cyclic sedimentation in the Permian Basin,* 2nd ed.: Midland, West Texas Geological Society, p. 203–232.

Meloy, D. U., 1974, *Depositional history of the Silurian carbonate bank of the northern Michigan Basin:* unpublished M.S. thesis, University of Michigan, 78 p.

Mesolella, K. J., Robinson, J. D., McCormick, L. A., and Ormiston, A. R., 1974, Cyclic deposition of Silurian carbonates and evaporites in the Michigan Basin: *Bull. Am. Assoc. Pet. Geol.,* v. 58, p. 34–62.

Newell, N. D., Rigby, J. K., Fischer, A. G., Whiteman, A. J., Hickox, J. E., and Bradley, J. S., 1953, *The Permian reef complex of the Guadalupe Mountains region, Texas and New Mexico: a study in paleoecology:* San Francisco, W. H. Freeman & Co., 236 p.

Newell, N. D., Imbrie, J., Purdy, E. G., and Thurber, D. T., 1959, Organism communities and bottom facies, Great Bahama Banks: *Bull. Am. Mus. Nat. Hist.,* v. 117, p. 117–228.

Nurmi, R. D., 1975, *Stratigraphy and sedimentology of the Lower Salina Group (Upper Silurian) in the Michigan Basin:* unpublished Ph.D. thesis, Rensselaer Polytechnic Institute, 261 p.

Nurmi, R. D., and Friedman, G. M., 1977, Sedimentology and depositional environments of basin-center evaporites, Lower Salina Group (Upper Silurian), Michigan Basin, p. 23–52, in J. H. Fischer, Ed., *Reef and evaporite concepts and depositional models:* American Association of Petroleum Geologists, Studies in Geology No. 5, 196 p.

Ottman, R. D., Keyes, P. L., and Ziegler, M. A., 1976, Jay field, Florida—a Jurassic stratigraphic trap, p. 276–286, in J. Braunstein, Ed., *North American oil and gas fields:* Tulsa, Okla., American Association of Petroleum Geologists, Mem. 24, 360 p.

Petta, T. J., 1980, Silurian pinnacle reef diagenesis—northern Michigan, p. 32–42, in R. B. Halley, and R. G. Loucks, Eds., *Carbonate reservoir rocks, notes for SEPM core workshop No. 1:* Tulsa, Okla., Society of Economic Paleontologists and Mineralogists, 183 p.

Playford, P. E., 1980, Devonian "Great Barrier Reef" of Canning Basin, Western Australia: *Bull. Am. Assoc. Pet. Geol.,* v. 64, p. 814–840.

Playford, P. E., and Lowry, D. C., 1966, Devonian reef complexes of the Canning Basin, Western Australia: *Bull. West. Austr. Geol. Surv.,* v. 118, 150 p.

Purdy, E. G., 1963, Recent calcium carbonate facies of the Great Bahama Bank. 2. Sedimentary facies: *J. Geol.,* v. 71, p. 472–497.

Purser, B. H., and Loreau, J. P., 1972, Structures sedimentaries et diagenetique precoces dans les calcaires bathoniens de la Bourgogne: *Bull. Fr., Bur. Rech. Geol. Minieres,* (Ser. 2), Sect. 4, No. 2, p. 19–38.

Reading, H. G., Ed., 1978, *Sedimentary environments and facies:* Oxford, Blackwell Scientific Publishers, 557 p.

Roehl, P. O., 1967, Stony Mountain (Ordovician) and Interlake (Silurian) facies analogues of Recent low-energy marine and subaerial carbonates, Bahamas: *Bull. Am. Assoc. Pet. Geol.,* v. 51, p. 1979–2032.

Ruzyla, K., 1980, The relationship of diagenesis to porosity development and pore geometry in the Red River Formation (Upper Ordovician), Cabin Creek Field, Montana: unpublished Ph.D. thesis, Rensselaer Polytechnic Institute, 200 p.

Ruzyla, K., and Friedman, G. M., 1980, Mechanisms controlling porosity in the Red River (Upper Ordovician) carbonate reservoir, Cabin Creek Field, Montana, p. 615–636, in P. O. Roehl, and P. W. Choquette, Eds., *Carbonate petroleum reservoirs: a case book* (preprint): New York, Springer-Verlag, in press, 719 p.

Ruzyla, K., and Friedman, G. M., 1981, Geological heterogeneities important to future enhanced recovery in carbonate reservoirs of Upper Ordovician Red River Formation at Cabin Creek Field, Montana:

SPE/DOE Second Joint Symposium on Enhanced Oil Recovery of the Society of Petroleum Engineers, p. 403–418.

Sanford, B. V., 1969, Silurian of southwestern Ontario: Ontario Petroleum Institute, 8th Annual Conference, p. 1–44.

Schmidt, V., 1977, Inorganic and organic reef growth and subsequent diagenesis in the Permian Capitan reef complex, Guadalupe Mountains, Texas, New Mexico, p. 93–132, in E. M. Hilman, and S. J. Mazzullo, Eds., *Upper Guadalupe facies, Permian reef complex, Guadalupe Mountains, New Mexico and Texas:* Midland, Texas, Society of Econ. Paleontologists and Mineralogists, Permian Basin Section, Pub. 77–16, 508 p.

Sears, S. O., and Lucia, F. J., 1979, Reef-growth model for Silurian pinnacle reefs, northern Michigan reef trend: *Geology,* v. 7, p. 299–302.

Shaver, R. H., Doheny, E. J., Droste, J. B., Lazor, J. D., Orr, R. W., Pollack, C. A., and Rexford, C. B., 1971, Silurian and Devonian stratigraphy of the Michigan Basin: a view from the southwest flank: Michigan Basin Geological Society Annual Field Excursion, Guidebook, p. 37–60.

Shaver, R. H., 1977, Silurian reef geometry—new dimension to explore: *J. Sediment. Pet.,* v. 47, p. 1409–1424.

Sigma Seismic Surveys Inc., 1977, *"Kamlot" seislog field study:* Houston, Texas, 64 p.

Sigsby, R. J., 1976, Paleoenvironmental analysis of the Big Escambia Creek-Jay-Blackjack Creek field area: *Trans. Gulf Coast Assoc. Geol. Soc.,* v. 26, p. 258–278.

Silver, B. A., and Todd, R. G., 1969, Permian cyclic strata, northern Midland and Delaware Basins, west Texas and southeastern New Mexico: *Bull. Am. Assoc. Pet. Geol.,* v. 53, p. 2223–2251.

Stoudt, E. L., 1979, Smackover Formation of the Gulf Coast region: manuscript, unpaginated, handout at AAPG-SEPM meeting.

Tyrrell, W. W., Jr., 1969, Criteria useful in interpreting environments of unlike but time-equivalent carbonate units (Tansil-Capitan-Lamar), Capitan reef complex, west Texas and New Mexico, p. 80–97, in G. M. Friedman, Ed., *Depositional environments in carbonate rocks:* Tulsa, Okla., Society of Economic Paleontologists and Mineralogists, Spec. Pub. No. 14, 209 p.

Ulteig, J. R., 1964, Upper Niagaran and Cayugan stratigraphy of northeastern Ohio and adjacent areas: *State of Ohio, Dept. of Natural Resources, Div. of Geol. Survey,* Rept. Inv. No. 51, 48 p.

Wilson, H. H., 1975, Time of hydrocarbon expulsion, paradox for geologists and geochemists: *Bull. Am. Assoc. Pet. Geol.,* v. 59, p. 69–84.

Wilson, H. H., 1977, "Frozen-in" hydrocarbon accumulations or diagenetic traps—exploration targets: *Bull. Am. Assoc. Pet. Geol.,* v. 61, p. 483–491.

Wilson, J. L., 1975, *Carbonate facies in geologic history:* New York, Springer-Verlag, 471 p.

Zenger, D. H., Dunham, J. B., and Ethington, R. L., Eds., 1980, *Concepts and models of dolomitization:* Tulsa, Okla., Society of Economic Paleontologists and Mineralogists, Spec. Pub. No. 28, 320 p.

Index

Abo reef, 192, 195,197
abyssal zone, 6
acetate peel, 2
acidification, 60
algae, 13, 61, 69, 149, 163, 165, 167, 192
algal mats, 6, 7, 14, 47, 61, 69, 77, 197
allochems, 118
alluvial placer deposits, 3
ammonites, 130, 134
anhydrite, 49, 78, 103, 111, 145, 149, 152, 163, 192, 197
annelids, 134
anticline, 203, 206
aphotic zone, 6
argillaceous limestone, 28, 101, 111, 115, 122, 130, 142, 160, 180, 195, 199
Asmari Formation, 17
atolls, 147, 176

bafflestone, 149, 151
barrier, 3, 4, 13, 18, 122, 123, 165, 167, 171
    back, 120, 122, 123
    oölitic, 28, 124, 127, 183, 186
barrier bars, 3, 6, 27
barrier reef, 64, 157, 160, 173–190, 191
basin, 3, 4, 23, 26, 39, 64, 178, 179, 180–181, 192
bathyal zone, 6
beach, 6, 18, 188
beachrock, 78, 129
bedding plane, 1
belemnites, 134
bindstone, 151
bioaccumulations, 176, 180
bioclastic limestone, 46, 101, 124, 143
bioherm, 142, 146, 162, 176
biomicrite, 103
biosparite, 55, 56, 142
biostrome, 165, 175, 176, 182, 187
biota, 5, 193
bioturbation, 69
bird's-eye fabric, 61, 77, 79
bivalves, 55, 56, 62, 64, 74
Black Lake field, 143–145, 146
block diagrams, 21, 23, 32, 33, 34, 188, 190

boundstone, 14, 47, 60, 163, 193
brachiopods, 103, 163, 175
breccia, 61, 69, 71, 77, 78, 86, 175, 180
    solution-collapse, 103, 133
brecciation, 7, 69
brecciola, 198, 199, 203, 205
brine ponds, 5
bryozoans, 56, 61, 103, 133, 142, 149, 162, 163

Cabin Creek field, 101
calcarenite, 142, 146, 180, 199, 204
calcrete, 163
calpionellids, 134, 138
capillary pressure curves, 41, 42–43
Capitan Limestone, 192, 193
casts, 194, 199
Cato San Andres field, 195, 196, 198
channels, 78, 178, 180, 203
chert, 79, 195, 199
coal, 3
coarsening-upward sequence, 15, 18, 19, 64, 124, 127, 199
coccoliths, 45, 199
concretions, 175, 178
coprolites, 133
coquinas, 135, 175, 176, 180
corals, 6, 13, 48, 52, 61, 64, 77, 133, 149, 151, 162, 163, 182
cores, 40, 152
crinoids, 56, 133, 151, 162, 163
cross-bedded, 103
cryptalgal laminites, 118
cryptocrystalline cement, 14, 45, 61, 69, 71, 78
cyclothems, 188

dedolomitization, 77
Delaware Basin, 190–194, 195, 196
delta, 3, 6, 53
depocenter, 23
desiccation, 7, 61, 69, 83
diagenesis, 2, 6, 7, 26, 39, 40, 46, 49, 50, 53, 55–58, 65–79, 87, 89, 90, 91, 117, 129, 152–155
    see also eogenesis; mesogenesis; telogenesis

diagenetic trap, 206–207
diastem, 14
disconformity, 14
dissolution, 7, 50, 53, 65, 69, 71, 77, 78, 171, 204
Dogger, 46, 47, 109, 119, 122, 123, 124, 127, 129
dogtooth spar, 51, 71, 78, 173
dolomicrite, 49, 111, 133
dolomicrosparite, 49, 53, 56
dolomite, 6, 49, 52, 54, 61, 71, 74, 78, 83, 87,
    101, 103, 105, 111, 113, 143, 149, 152, 163,
    167, 173, 195
dolomitization, 52, 53, 54, 62, 64, 74, 77, 78, 79,
    83, 101, 103–107, 117, 140, 151, 152, 163
  epigenetic, 74
  syngenetic, 101, 103, 107
dolosparite, 111, 114
dolostone, 62, 71, 106, 143
drusy spar, 51, 65
dunes, 3
Dunham's classification, 14, 150, 151

echinoderms, 44, 77, 78, 103, 129–130, 133, 142,
    165, 169
electrical resistivity, 59
Empire Abo field, 194, 195
eogenesis, 69–74, 89
epeiric sea, 4, 6, 101
epigenesis, 152
epitaxial overgrowth, 69, 74, 77, 151, 152
equant calcite, 71, 145
eustacy, 19, 23, 26
evaporites, 6, 7, 49, 62, 71, 79, 101, 103, 108, 111,
    118, 149, 157, 158, 160, 161, 163, 192

facies models, 1, 2, 100
fans, deep-sea, 6
faults, 78, 173, 203, 205
fenestral fabric, 61, 77, 79, 83, 120, 133
fining-upward sequence, 15, 18, 19, 64, 199
fissure, 41, 71, 74, 77, 153–155
floatstone, 151
foraminifera, 13, 45, 47, 53, 61, 77, 115, 165, 167,
    182, 199
framestone, 151
freshwater, 65, 68, 71, 74, 78, 101, 163

gamma-ray log, 41, 59, 105, 108, 123, 167
gastropod, 45, 103, 167
geopetal fabric, 175
glaciers, 3
graded bedding, 103, 180
grainstone, 14, 15, 42, 77, 116, 133, 165
gravitational cement, 7
gypsum, 78, 192

hadal zone, 6
halite, 78, 163, 192

hardground, 14
high-Mg calcite, 69, 71, 78, 129
hydrocarbon, 40, 59, 141, 143, 146, 147, 194,
    206–207
hypidiotopic, 103

intertidal, 5, 6, 14, 53, 71, 77, 111, 113, 163, 167
intraclasts, 61, 86, 103, 197
isostacy, 26

joints, 78

kaolinite, 145
karsts, 3, 58, 79
Kemnitz field, 195

lagoons, 3, 28, 78, 122, 123, 127, 151, 179
laterologs, 108
leaching, 62, 64, 83, 103, 118, 160, 163
Lennard Shelf, 173–190
levees, 3, 6, 188
lime mud, 7, 13, 14, 46, 61, 69, 108, 127
lithification, 69, 77
lithofacies, 130–133, 143–145
lithology, 14, 17, 59, 108, 109, 114, 116–117, 120,
    167, 168
Lons Formation, 111
low-Mg calcite, 71, 78

mangroves, 6
Mano Formation, 111
marine phreatic, 65
marls, 28, 111, 122, 127
megasequence, 18, 24, 25, 127, 135, 203
meniscus cement, 7, 78
mercury injection, 41
mesogenesis, 68, 74–77
Michigan Basin, 157–165
micrite, 14, 57, 71, 111, 115, 118, 141, 149, 163,
    173
micritization, 14, 74, 79
microaggregates, 127
microcrystalline limestone, 122, 124
microfacies, 108, 110
microfilaments, 134
micropaleontology, 2
micropeloidal cement, 55, 69, 127–129
microstratigraphy, 130
miliolids, 142, 167
millidarcy, 39, 40
Mishrif Formation, 165–173
molluscs, 61, 103, 142
mudstones, 15, 48, 115, 135, 151, 163, 167, 180,
    192

nautiloids, 175, 177, 180
neomorphic, 65, 79
neptunian dikes, 178, 188
neutron logs, 59, 167

oncolites, 103, 120, 143, 175, 176
oöids, 6, 44, 49, 53, 55, 56, 61, 62, 78, 103, 118, 120, 127, 129, 133, 175, 180, 187
oölitic limestone, 28, 116, 120, 124, 142, 145, 146, 206
oömicrite, 142
oösparite, 142
orbitoidids, 142, 165
ostracodes, 103, 142, 192
oyster banks, 61

packstones, 15, 42, 47, 50, 53, 133, 142, 143, 165
paleoshoreline, 117, 142, 145
palynology, 2
Paris Basin, 44, 46, 47, 51, 55, 56, 119−130
peat, 3
pedogenesis, 77
pelagic oozes, 151
pelagic organisms, 13, 20, 176, 180, 199, 205
pellets, 120
peloidal limestone, 124
peloids, 45, 103, 115, 127, 175, 177
peritidal, 5, 13, 101, 107, 112
permeability, 39, 40−41, 42, 59, 87, 93, 97, 129, 130, 143, 145, 146, 151, 153, 156, 170
phreatic zone, 65, 68, 78, 118
photic zone, 3, 6
piedmonts, 3
pinnacle reef, 147, 157−165
pisolites, 71, 142, 163
plug, 40
porosity, 17, 39−64, 72−73, 77−78, 82−99, 120, 129, 143, 145, 146, 151, 152, 156, 167, 170
   breccia, 44, 60, 61, 62−63, 86
   fenestral, 45, 60, 61, 64, 69, 86
   fracture, 60, 62−63, 83, 86, 87, 197, 205
   framework, 44, 60−61, 64, 86, 87
   intercrystalline, 52, 60, 61−62, 71, 74, 83, 86, 87, 101, 107, 118, 163, 195, 197
   interparticle, 44, 51, 52, 55, 60, 61, 64, 65, 69, 83, 86, 87, 107, 127, 171, 204, 205
   interskeletal, 60
   moldic, 53, 60, 62, 71, 86, 107, 197
   primary, 60−61, 64, 65, 66, 69, 77, 83, 100−101, 204, 205
   secondary, 39, 48, 49, 60, 61−63, 64, 65, 83, 101, 204
   shelter, 45, 61, 86
   vug, 47, 48, 50, 51, 53, 60, 62, 77, 86, 107, 163, 195
   weathering, 60, 63
prealveolinids, 165
pressure solution, 39, 55, 56, 62, 74, 79, 130, 206
progradation, 18, 26

Rainbow reefs, 147−157
recrystallization, 74, 83, 96, 99, 151, 173
Red River Formation, 101−107
reef, 33, 52, 53, 78, 141, 143, 147−165, 186, 187,

192−193, 196
regressive sequence, 15, 26, 35, 64, 124, 135, 165, 168, 169−170
rim cement, 56, 77, 129, 130, 143, 145, 151, 204
rudistid banks, 61, 167
rudstone, 151

sabkha, 5, 14, 17
saddle dolomite, 145
salinity, 3, 6, 7, 14, 59, 116
sandstones, 39, 135, 141, 173
Scaglia Calcaire, 198−206
scanning-electron microscope, 2, 59
scour surface, 14
serpulid worms, 142
shales, 15, 59, 141
shelf, 3−6, 13, 54, 55, 64, 130, 160, 165−173, 178, 179−180, 192
shoal, 3, 4, 51, 55, 56, 138, 141, 167, 169
silica, 8, 11, 12
siliciclastic, 61, 140
Sligo Formation, 140−146
slope, 6, 33, 64, 151, 179, 180−181
slope deposits, 101, 117, 179, 180−181
slump, 6; 26, 175, 177, 193
Smackover Formation, 113−118
solution, 41, 78, 87, 143, 205
spontaneous potential (S.P.), 59, 122
stromatactis, 149, 178
stromatolites, 162, 163, 175
stromatoporoids, 142, 151, 162, 163
stylolites, 55, 56, 63, 74, 130, 167, 171, 182
subduction, 26
subsidence, 16, 19, 23, 26, 35
subtidal, 5, 6, 64, 163, 167
supratidal, 4, 5, 6, 14, 64, 71, 77, 103, 163, 167
synsedimentary, 145, 188
syntaxial overgrowth, 145

talus, 6, 151, 193, 205
telogenesis, 68, 75, 77, 89
tidal flats, 3
transgression, 26, 124, 138, 143, 183
transgressive sequence, 15, 18, 19, 22, 26, 35, 64, 127, 135
trilobites, 103, 177
turbidites, 6, 18, 198−206
turbidity current, 18, 198, 205

uncomformity, 14, 165, 169, 171

vadose zone, 65, 68, 71, 78, 79, 118

wackestone, 15, 42, 47, 143, 163, 169, 192, 199
Walker Creek field, 117−118
wettability, 41
Williston Basin, 101−107

xenotopic, 103